BASIC *Mapwork* s

Simon Ross

Text © Simon Ross 2003

Original illustrations © Nelson Thornes Ltd 2003

The right of Simon Ross to be identified as author of this work has been asserted by him in accordance with the Copyright, Designs and Patents Act 1988.

All rights reserved. No part of this publication may be reproduced or transmitted in any form or by any means, electronic or mechanical, including photocopy, recording or any information storage and retrieval system, without permission in writing from the publisher or under licence from the Copyright Licensing Agency Limited, of 90 Tottenham Court Road, London W1T 4LP.

Any person who commits any unauthorised act in relation to this publication may be liable to criminal prosecution and civil claims for damages.

Published in 2003 by:
Nelson Thornes Ltd
Delta Place
27 Bath Road
CHELTENHAM
GL53 7TH
United Kingdom

03 04 05 06 / 10 9 8 7 6 5 4 3 2 1

A catalogue record for this book is available from the British Library

ISBN 0–7487–7409–2

Page make-up and illustration by DC Graphic Design Ltd

Artwork 9.3 by Richard Morris

Printed in Croatia by Zrinski

Cover photograph: Pend Oreille River surrounded by forested hills and farmland, Eastern Washington, USA.
Photograph on title page: San Francisco Bay Area, California, USA.

Contents

Acknowledgements	4
Preface	5
Location of places in the Mapskills Units	6

'How to ...'

SECTION A: Interpreting maps	7
SECTION B: Interpreting photographs	12
SECTION C: Presentation of data	14
SECTION D: Looking for patterns	16

Mapskills Units

1	Introduction to using maps and photographs: The Millennium Stadium, Cardiff	19
2	Road map: Northern England	24
3	City: Hulme, Manchester	26
4	Town: Buxton, Derbyshire	32
5	Land use change: Shoreham-by-Sea, Sussex	35
6	Village: Castleton, Derbyshire	40
7	Limestone scenery: Ingleborough, North Yorkshire	44
8	River flood: Malton, North Yorkshire	49
9	Coastal erosion: The Green Bridge of Wales, Pembroke	52
10	Tourism: The Eden Project, Cornwall	56
11	Rivers (1): The Rhône, Switzerland	60
12	Rivers (2): The Rhône, France	64
13	Rivers (3): The Rhône delta, Camargue, France	68
14	Climate: Comparing climates in Europe	74
15	Living in the mountains: Chamonix, French Alps	77
16	Industry: BMW car plant at Dingolfing, Germany	80
17	Quarrying: Glensanda 'superquarry', Scotland	83
18	Volcano: Mount Etna, Italy	86
19	Landslide: Karakoram Highway, Pakistan	90
20	Population: The Earth at night	94
21	Tourism: The Great Barrier Reef, Australia	96
22	Water supply: Las Vegas, Nevada, USA	99
23	LEDC city: Mexico City, Mexico	102
24	Tropical rainforest: Pasoh Forest Reserve, Malaysia	106

Key Stage 3 Assessment Tasks

A	Locating a new hypermarket: Aylesbury, Buckinghamshire	112
B	Living with earthquakes: California, USA	118

Map Keys

OS 1:50,000 Landranger map series	123
OS 1:25,000 Explorer map series	124
Land use maps	125–126
1:25,000 French survey maps	127–128

Acknowledgements

Author's note

The production of this book has been very much a team effort and I am indebted (again) to the good-humoured, professional and highly efficient support that I have received from Barry Page, Vicky Gannicott, Penni Bickle, Nick Bullmore, and Katherine James. Whilst it has been hard work, it has been great fun.

I am also extremely grateful to the following people for their help and assistance:

Aidan O'Rourke

Barbara Brailey (BMW)

Charles Eaton

Dr S. M. Tilling (Field Studies Council)

Foster-Yeoman

John Edwards

Joseph Kerski of the United States Geological Survey

Karen Frenkel

Miss D. D. Bird

Mr P. Ossum

Peter Smith

Tony Waltham

Finally, I should like to thank my wife Nikki, children Susannah and James, and cats Gus, Charlie and Kim for putting up with me tapping away in my study when I should have been cleaning the bath, washing the cat, making the dinner, playing with Barbie dolls (!) and kicking a football.

Dedication

To all my family and, on this occasion, with particular gratitude to Russ for many years of guidance and wisdom throughout the ups and downs of family life.

Copyright material

The author and publishers are grateful to the following for permission to reproduce copyright mapping and photographs.

References are to figure numbers/letters within each case study. For example, 24.2 refers to Figure 2 in case study number 24, and A3 refers to Figure 3 in Assessment Task A.

Photographs

Associated Press/AP: 18.2, B5.1, B5.2; Associated Press/NASA: 18.1; BMW: 16.1; ClockTowerWeb: H11; Collections/Ashley Cooper: 7.4; Collections/Robin Weaver: 7.5; Corbis/Raymond Gehman: 14.3 (BL); Eden Project: 10.2; John Edwards: A3; English Heritage Photo Library: 6.1; Eye Ubiquitous/John Miles: 21.1; Field Studies Council/Steve Tilling: 24.2, 24.5, 24.6, 24.7; FLPA/Mark Newman: 22.3;

Foster Yeoman: 17.3; Karen Frenkel: 6.2; Getty/Allsport: 1.1; Getty Images/Clive Brunskill: 1.7; Getty: 15.3, 15.4; James Davis: 21.3; John Cleare Mountain Camera: 11.2; Manchester City Library/Simmons Aerofilms: 3.2; Manchester City Library: 3.3; Aidan O'Rourke: 3.5; Peter Smith Photography: 8.3; Rowland Aerial Photography: 5.3; South American Pictures: 23.1, 23.2; Science Photo Library: H13, 12.1; Science Photo Library/M-Sat Ltd: 2.1; Science Photo Library/David Parker: 10.1; Science Photo Library/ESA,Eurimage: 13.2; Science Photo Library/Simon Fraser: 14.3 (TR); Science Photo Library/W T Sullivan: 20.1; Sealand Aerial Photography: H12; Still Pictures/Nigel Dickinson: 13.5; The Photolibrary Wales: 9.1; Tony Waltham/Geophotos: 19.1, 19.3; Trip/B Turner: 13.4; Trip/M Feeney: 14.3 (TL); Simon Warner: 4.1, 7.2, 7.6; www.jasonhawkes.com: 1.5, 4.2, 13.3, 14.3 (BR), 17.1; York & County Press: 8.1.

Cover photo: Digital Vision WA(NT)/Sunset Avenue Productions.

Title page photo: Digital Vision 7(NT).

Mapping

AA/Ordnance Survey: 2.2.

Association of Bay Area Governments: B7

Topographic Map 1:50.000 sheet Nr. L 7340, reproduction with permission by the Bavarian Survey-Office Munich, Nr. 1930/03: 16.2.

GeoCenter: 23.3i & ii.

hemamaps.com.au: 21.2.

© MICHELIN 2003 extract from Map No. 523, 1st edition Authorisation No. 0306259: Front cover (right), 15.2.

National Imagery and Mapping Agency, www.nima.mil: 19.2, 24.3.

This product includes mapping data licensed from Ordnance Survey® with permission of the Controller of Her Majesty's Stationery Office, © Crown Copyright. All rights reserved. Licence No. 100017284: Front cover (left & middle), HA, 1.3, 1.4, 3.4, 4.3, 5.1, 5.2, 6.3, 7.3, 8.4, 9.2, 10.4, 17.2, keys to OS mapping pp123 & 124.

swisstopo – reproduced by permission of swisstopo (BAO35355): 11.3

Touring Club Italiano: 18.3

United States Geological Survey: B6

Universal MAP: 22.2

www.ignfi.fr: 15.5

Every effort has been made to contact copyright holders. The publishers apologise to anyone whose rights have been inadvertently overlooked, and would be happy to rectify any errors or omissions.

Preface

Basic Mapwork Skills is intended to provide a support for the teaching of practical map skills and photograph interpretation, which is a central component in the study of geography. Particular consideration has been given to providing resources appropriate for Key Stage 3, although maps and photographs can of course be used at any level in schools.

Basic Mapwork Skills has the following features:

- Large full-page photographs and maps.
- OS maps at 1:5,000, 1:10,000, 1:25,000 and 1:50,000 scales.
- A selection of foreign maps at different scales.
- Satellite, aerial and ground photographs.
- A variety of map types, including land use maps and town plans.
- A written introduction to each Mapskills Unit.
- A range of suggested activities, addressing key aspects of practical mapskills and photograph interpretation as outlined in the National Curriculum.

- An introductory 'How to…' section which covers the key practical skills, to act as a reference for students as they tackle the practical work within each Mapskills Unit.
- References for further information using the internet.
- Specially commissioned Key Stage 3 Assessment Tasks with comprehensive level descriptions.

Basic Mapwork Skills is a book for dipping into and for using as a resource in teaching and learning practical skills in geography. Students can also use and expand the material on the localities in the Mapskills section, in order to develop their own case studies, particularly if they follow up the weblinks.

Through its use of large photographs and map extracts, it is hoped that Basic Mapwork Skills will enthuse and inspire future generations of geographers.

The publisher has endeavoured to ensure that the URLs (website addresses) in this text are correct and active at the time of going to press. However, the publisher takes no responsibility for the websites, and cannot guarantee that a site will remain live or that the content is or will remain appropriate.

Location of places in the Mapskills Units

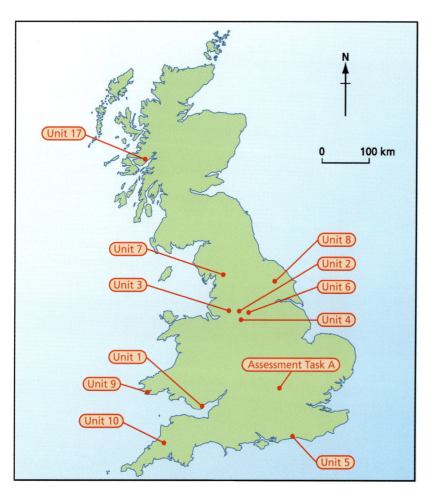

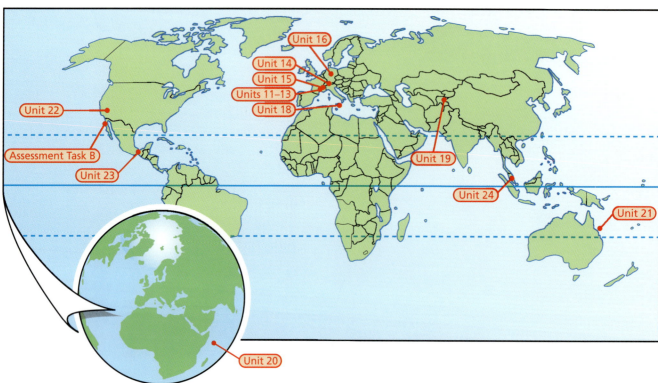

'How to...'

SECTION A: Interpreting maps

1 What is a map?

A **map** is a drawing showing a bird's-eye view of an area. Maps come in all sorts of shapes and sizes. Some show small areas, such as shopping centres or school campuses – these are often called **plans**.

Most maps cover large areas, often many kilometres. Maps can be bought as single sheets or as collections in an atlas. Atlases contain many maps covering the whole world and some may study themes, such as climate and population, as well as the basic geography of countries. A road atlas contains maps that are useful for drivers trying to find the best route between places.

In the UK many people use maps produced by the **Ordnance Survey**. These maps, usually known as **OS** maps, show lots of detail about an area including the roads, the buildings, countryside features and the physical landscape. Look at the map of Maidenhead on page 18 to see the how much information is presented.

2 What is scale?

A good and accurate map must have a scale. This shows how distance on the map (in cm or mm) relates to real-life distance on the ground.

On a map, scale is shown in two ways (see Figure 1).

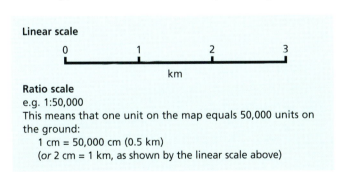

1 *Scale*

3 What are map symbols?

Maps contain a huge amount of information. This is made possible by using **symbols** instead of written labels, which would take up far too much space. Many symbols are clear in their meaning, but they are always

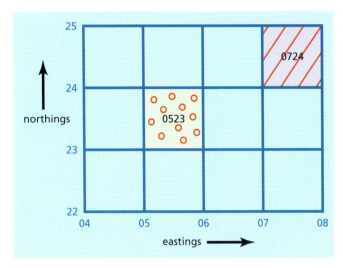

2 *How to find a four-figure grid reference*

explained in a **key**. The key is usually found at the base or to the side of a map.

OS maps use a large number of symbols because they tell us a great deal about an area. Look at the Maidenhead map on page 18. Notice that the symbols on the map are explained in a key at the end of the book, on page 123.

4 How to find grid references

Look at the Maidenhead map on page 18. Notice that it has been divided into boxes using light blue lines, and that each blue line has a number written by it. The numbers increase in two directions:

- from left to right across the map (these lines are called **eastings**)
- from bottom to top across the map (these lines are called **northings**).

The grid system is very useful as it makes it possible to locate places using **grid references**. There are two types of grid reference:

- **four-figure grid references**, which are used to identify a particular grid square
- **six-figure grid references**, which can locate places more accurately within a grid square.

How to find a four-figure grid reference

The main thing to remember is that each gridline number refers to the next square, either across the map or up the map. To locate a particular grid square, you first read along the bottom to find the eastings number. Then you read up the side to find the northings number. The two sets of numbers give the four-figure grid reference. Look at Figure 2 to see how it works.

'How to ...'

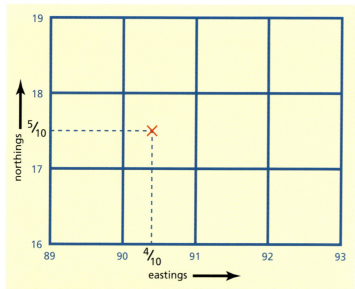

Looking at the eastings values, X lies 4/10 of the way between 90 and 91. This is written as **904** and makes up the first three figures of the six-figure reference. Looking at the northings values, X lies 5/10 between 17 and 18. This is written as **175** and makes up the second set of figures.

The full six-figure reference for X is **904175**.

3 *How to find a six-figure grid reference*

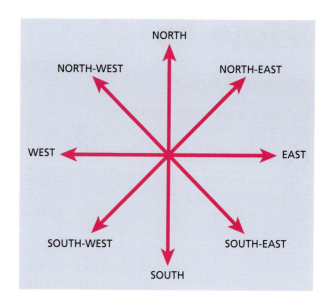

4 *Compass directions*

Now turn to the Maidenhead map on page 18. Find the major motorway junction towards the bottom of the map. Its four-figure reference is 8878. Now see if you can find out which village is in grid square 9079. It should be the village of Bray!

How to find a six-figure grid reference

To find a six-figure grid reference, you have to imagine that the distance between two gridlines is divided into tenths. Whilst it is possible to use a ruler to split a square into tenths, it is good enough to make a careful estimate.

Look at Figure 3 to see how this works and how a six-figure grid reference can be given.

Now look at the Maidenhead map (page 18). Find the big motorway junction again. Notice that its six-figure grid reference is 887785. Now find the village of White Waltham in the bottom left-hand corner of the map. There is a 'place of worship with a tower' (study the key on page 123 to find the symbol used) in the village. What is its six-figure grid reference? Would you agree that it is 855775?

The most important thing to remember with six-figure grid references is that you must always give the eastings values first, followed by the northings values, just as you do for four-figure references. Some people use the saying *along the corridor and up the stairs* to help them to remember which way round to write the numbers!

5 How to give compass directions

It is very useful to describe where one place is in relation to another. To do this we use compass directions (see Figure 4).

The most important thing about giving a compass direction is to state very clearly which way you are looking or travelling. For example, on the Maidenhead map (page 18), Holyport is to the **east** of White Waltham whereas White Waltham is to the **west** of Holyport! If you wanted to go from White Waltham to Woodlands Park, in which direction would you travel? The answer, in case you are not sure, is 'north-east'.

6 How to measure distance

To measure distance, you need to convert the distance between two places on the map (usually in cm) and then convert it using the scale to give you the real-life distance on the ground. The easiest way to do this is to use a ruler to find out the number of centimetres between the two places and then place the ruler alongside the linear scale to convert the measurement to kilometres (see Figure 5).

Turn to the Maidenhead map (page 18). What is the straight-line distance between the 'place of worship with a tower' in White Waltham and the large motorway junction at 887785? You should find that it is just less than 3.5 km.

Measuring curved distances, say along a road or river, is rather more time-consuming and fiddly. The simplest way of measuring a curved distance involves pivoting the

'How to ...'

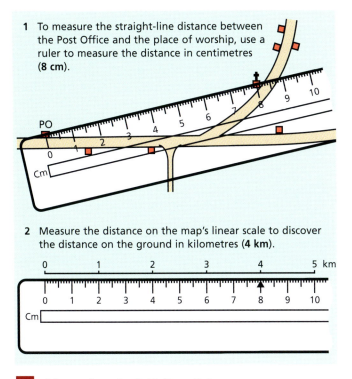

5 *Measuring straight-line distance*

straight edge of a piece of paper alongside the curved line (see Figure 6), marking off straight sections on the paper's edge. The total straight-line distance along the edge of the paper can then be converted into kilometres using the linear scale.

Now try to find the curved-line distance along the M4 motorway from where it joins the map at 864770 to where it leaves the map at 930801. You should find that this distance is about 7.5 km.

7 How to describe landscapes

Ordnance Survey maps tell us a lot about the 'ups and downs' (the **relief**) of a landscape. Turn to the Maidenhead map extract (page 18) and examine grid square 8483. Apart from the obvious main road passing through the square, notice that there are several rather faint brown lines. These lines are called **contours**.

Contours are lines that join points of equal height. They are usually drawn at 10 or 5 metre intervals, although it is important to check in the key. Notice in grid square 8483 that only one contour value is given – 80 m. This marks the top of a small hill, with contours of lower values around it. You often have to trace a contour across the map to find its value, or use neighbouring values to act as a guide.

The spacing between contours tells us how steep a slope is (see Figure 7). The closer together the contours are, the

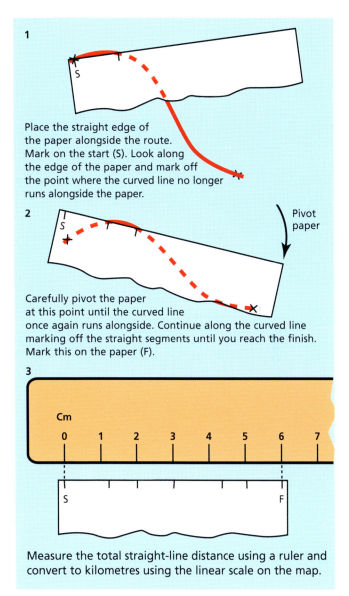

Measure the total straight-line distance using a ruler and convert to kilometres using the linear scale on the map.

6 *Measuring curved distance*

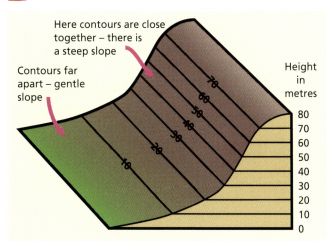

7 *Contours and slope steepness*

'How to ...'

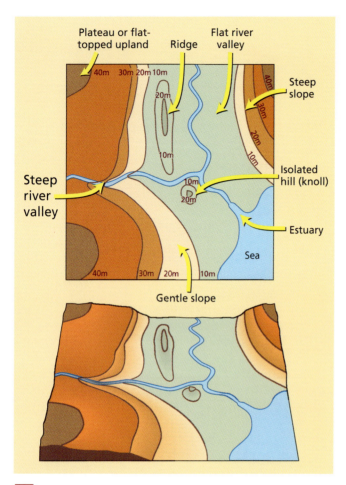

8 *Contours and landscapes*

steeper the slope. Contour patterns help to inform us about some of the features of the land, as Figure 8 illustrates.

The height of the land is shown in other ways too. **Spot heights** are small black dots with a number written alongside. The number is the height in metres above sea level. In grid square 8483 on the Maidenhead map there is a 52 m spot height next to the road. What is the spot height at 850826? You should discover that it is 27 m higher than the one in 8483!

There is more information about height in the keys on pages 123–124.

8 How to draw a cross-section

A cross-section is a 'slice' through the landscape. It helps us to see more accurately what the landscape looks like. To draw a cross-section, you need to follow the steps below and refer to Figure 9.

1. Draw a line of section (from A to B) across the contours (Figure 9a).
2. Study the patterns of the contours and try to imagine what is happening to the land. Is the land rising or falling? Is there a valley?
3. Place the straight edge of a piece of paper along the line of section (Figure 9b). Mark the start (A) and the finish (B). Now mark the points where each contour crosses the edge of the paper. Write the value of each contour on the paper. Check to see whether any contours are 'doubling-back'. Mark on any other important details such as rivers.
4. Now place the paper along a horizontal line of the same length as the line of section, on a piece of graph paper (Figure 9c).
5. Choose a vertical scale that shows the variations in height but does not make the landscape look unrealistic.
6. Carefully draw crosses at the correct height for each contour value (Figure 9c). To do this you need to read up from the bottom.
7. Use a freehand curve to join up the crosses (Figure 9d). Continue the curve to both axes. Notice how this has been done at the start of the section, where the value at A lies somewhere between 60 m and 70 m.
8. Complete your section by writing labels and giving your diagram a title (Figure 9d).

Drawing a cross-section is one of the most difficult geographical skills that you will attempt. However, with a bit of practice it isn't too difficult, and the result can be really impressive. Here are a few tips:

- Always use a sharp pencil and be prepared to rub out mistakes.
- Double-check that you have written down the correct contour values on your piece of paper.
- Take time to consider an appropriate vertical scale. The following scales tend to work well, although it will depend a bit on 'trial and error':

  1:50,000 map   1 cm = 100 m

  1:25,000 map   1 cm = 20 m

- If the contours are very close together, then mark every other one.

9 How to describe townscapes

Ordnance Survey maps tell us a lot about towns. Look at the Maidenhead map on page 18 to see how it is possible to learn a lot about townscapes.

- **Roads** Several main roads meet in the centre of Maidenhead. Perhaps this helps to explain why Maidenhead grew up to be a large town? Some roads are dual carriageways, for example part of the A308.

'How to ...'

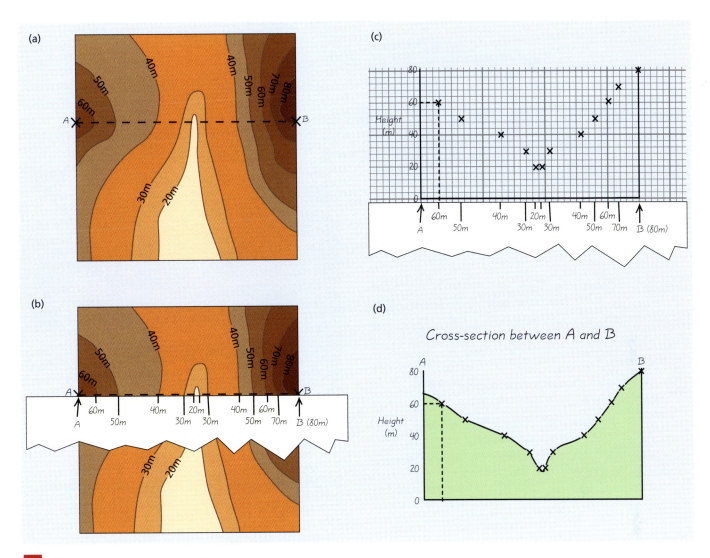

9 How to draw a cross-section

It is possible to see roundabouts on some of the roads. Minor roads are shown in yellow and local roads in white. See how some roads form straight-line patterns whereas others form curves. You can see modern housing estates (e.g. grid square 8682) with their curved roads, closes and culs-de-sac. There is a by-pass to the west of Maidenhead – the A404(M) and the A404.

- **Built-up area** The pale pink/brown colour shows the extent of the built-up area of Maidenhead. This is where the houses and shops are. The white spaces in between are areas of open ground, such as parks. Some important buildings, such as schools, are shown separately. Locate a school, giving its six-figure grid reference.
- **Functions** There are several functions and services shown on the map. In the town centre (8880 and 8881) you can see places of worship, the railway station, and the Town Hall. Can you find some other functions or services in Maidenhead?

10 How to draw a sketch map

A sketch map is a simple map that shows some but not all of the detail that exists on an Ordnance Survey map. Sketch maps are useful to us when we wish to study something on its own, like road networks or the location of forests.

The gridlines on a map make it possible to draw a fairly accurate sketch map – they are used as guidelines for transferring information. Sketch maps can be drawn to the same scale as the original map or they can be enlarged or reduced in size.

To draw a sketch map you need to follow the steps on the next page, and refer to Figure 10, which is a sketch map of part of Maidenhead. Remember to use pencil first!

'How to ...'

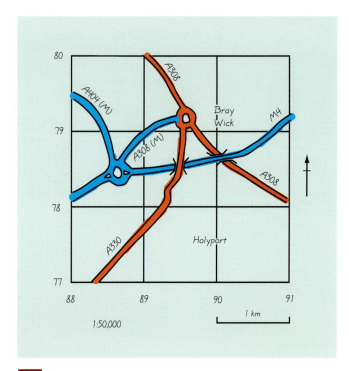

10 *Sketch map showing main roads south of Maidenhead*

1 Draw a grid frame exactly matching the area of the map to be sketched. If you are doing the sketch to the same scale as the original then the squares will need to be the same size. You may wish to reduce or enlarge the map by altering the size of the squares.
2 Write the grid numbers at the edge of your frame.
3 Now very carefully transfer the information that you want from the original map to your own, using the gridlines to guide you.
4 Complete your sketch by:
   • adding colours and a key
   • writing a title
   • adding place names and other labels
   • adding a scale and a north point.

SECTION B: Interpreting photographs

1 How to interpret ground photographs

Ground photographs are photographs taken by someone standing on the ground. They show what a place looks like as we would see it if we were standing on the ground.

To interpret a ground photograph you need to study it closely and look for clues to help you understand what is happening. For example, if trees are in leaf and people are wearing shorts then it is probably summer. If people are smiling and having fun it tells you that it is a nice place to be.

Figure 11 is a ground photograph taken in the centre of Maidenhead. Take time to study the photograph. What are the different shops? Are there many people in the photograph? Does it look like a place you would like to visit? Why?

11 *Centre of Maidenhead*

2 How to interpret aerial photographs

Aerial photographs give us much the same view of an area as we would see when looking out of an aeroplane window. **Vertical** aerial photographs look directly down on an area, much as a map does. **Oblique** aerial photographs (see Figure 12) look down on an area at an angle.

Aerial photographs are excellent in showing what an area looks like. They can help us to understand and bring to life the detail shown on a map.

Figure 12 is an oblique aerial photograph of Maidenhead. The circular tanks at the bottom of the photograph are part of a sewage works. You can find the sewage works on the Maidenhead map at grid reference 894804. The photograph was taken looking north over Maidenhead. Take time to locate on the map extract some of the features identified on the photograph.

To work out which direction a photograph is looking, you first need to locate on a map some of the features shown at the bottom, middle and top of the photograph. This gives you a line of sight. Then use the compass directions on the map to help you work out which way the photograph is looking.

'How to ...'

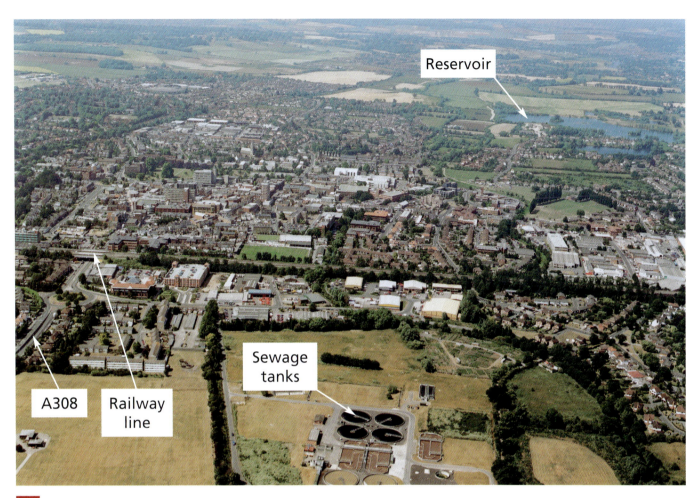

12 *Oblique aerial photograph of Maidenhead looking north*

3 How to interpret satellite photographs and images

Satellites can provide us with very accurate and detailed photographs, often covering large areas of the Earth's surface. Many modern maps are produced using satellite photographs because they are so accurate and up-to-date.

Computers can create satellite images that use false colours to help identify features of interest, for example green crops, surface water or settlements. Figure 13 is a satellite image of the Maidenhead area.

4 How to draw a sketch from a photograph

It is sometimes useful to draw a sketch from a photograph in order to show features of interest. Figure 14 is a sketch of part of Figure 12. To draw a sketch you should follow the steps on the next page, working in pencil as always.

13 *Satellite image of Maidenhead*

'How to ...'

1. Draw a frame to the same shape as the area on the photograph that you wish to sketch.
2. Draw some of the main 'lines' (such as field boundaries, roads, edges of uplands, etc.) in your frame. These will then act as guides to help you locate other features.
3. Continue to add the other details on the photograph that you are interested in. Be bold and clear and, as a general rule, avoid shading.
4. Now add any labels, such as place names or geographical features. Give your sketch a title.

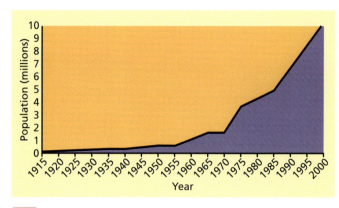

15 *Growth of Bogota, Colombia*

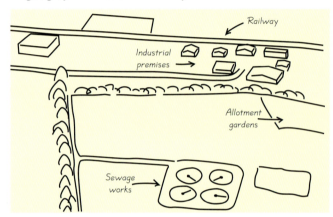

14 *Sketch of part of Maidenhead aerial photograph (Figure 12) looking north*

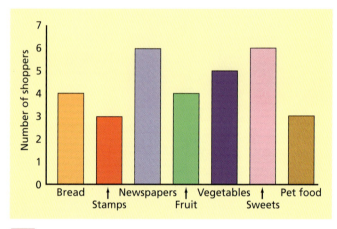

16 *Bar graph showing the number of items bought by shoppers at a local shopping centre*

SECTION C: Presentation of data

1 Drawing line graphs

A line graph is most commonly used to show changes over a period of time, for example the population growth of Bogota, capital of Colombia (Figure 15). The most important thing to remember when drawing a line graph is to make sure that you have equal time intervals along the horizontal (*x*) axis.

2 Drawing bar graphs and histograms

Bar graphs are drawn to show the frequency or amount of different things, such as items bought at a local shopping parade (Figure 16). Notice that there is data on the vertical axis only.

However, if the information is part of a single, continuous data set, then a histogram is the correct method to use (Figure 17). Here the bars are drawn directly alongside each other. Notice on Figure 17 that data is written along both of the graph axes and that it is continuous, going from low values to high values.

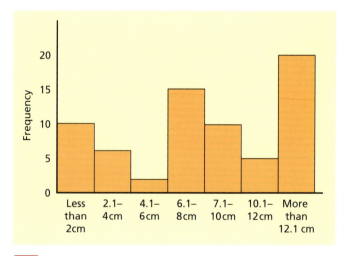

17 *Histogram showing the size distribution of pebbles measured in a Dartmoor stream*

'How to ...'

3 Drawing pie charts

A pie chart is a circle that has been divided into segments, rather like the slices of a pie. It is usually drawn to show the proportions of a total, for example the emissions of sulphur dioxide into the atmosphere (Figure 18).

To draw a pie chart you first need to draw a circle using a pair of compasses. Then you need to convert the raw data for each category into degrees before using a protractor to divide the 'pie' into segments. The segments can be coloured, but don't forget to add a key.

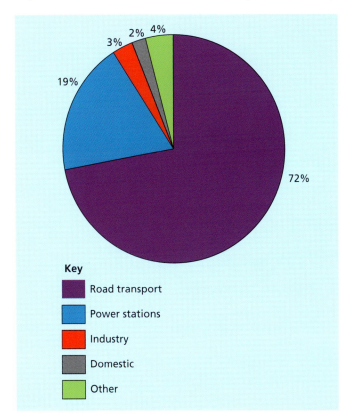

18 Sulphur dioxide emissions by sector

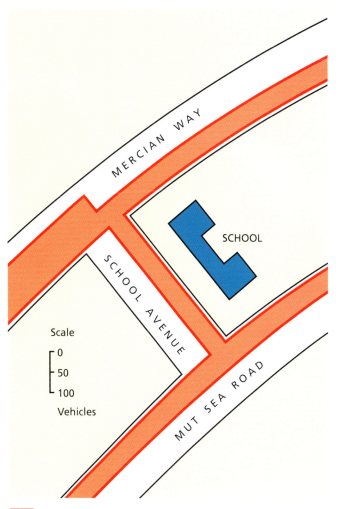

19 Traffic flow map 10.00–10.15 am, Monday 23 May

4 Drawing flow lines

Flow lines are a very good way to show movement between places. Figure 19 shows the movement of traffic in the roads near to a local school. Notice that the lines have been drawn with a thickness that represents the amount of traffic.

5 Drawing a choropleth map

A choropleth map uses different colours or density of shading to show the spread of data over an area. Figure 20 is a choropleth map that shows infant mortality rates in the world. Notice that the map has several important features that are typical of choropleth maps:

- Data is divided into a number of groups or categories.
- There are 5 categories. The ideal number of categories is between 4 and 8.
- None of the category values overlap.
- The colours become darker as the values increase.

15

'How to ...'

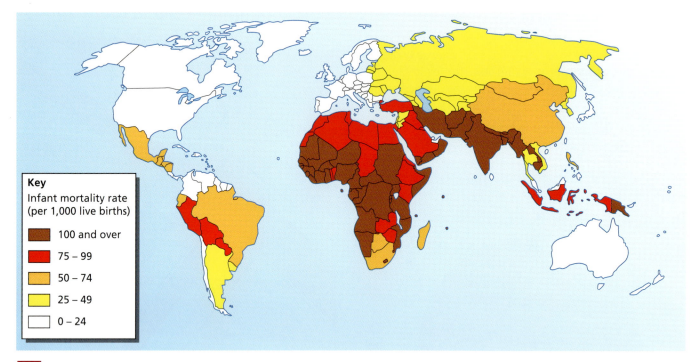

20 *World infant mortality rates per 1,000 live births, 1990*

6 Drawing a climate graph

A climate graph shows temperature and rainfall data for a place on a single diagram. Average monthly temperatures are shown as a line graph and average monthly rainfall as a histogram (Figure 21). If you are comparing climates for two or more places it is important to keep to the same scales on the graphs.

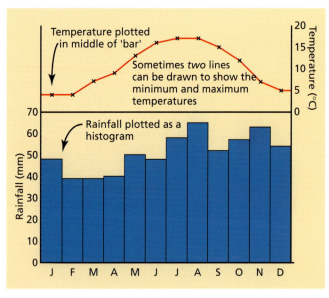

21 *A climate graph for London*

SECTION D:
Looking for patterns

1 How to 'describe'

To describe a map, photograph or diagram you need to put into words what it shows. If possible refer to the information available to you, which might include places or data. Be as precise and as detailed as you can.

A description of Figure 20 might begin:

The highest rates of infant mortality (100 and over per 1,000 live births) are found in Africa, the Middle East and parts of SE Asia. The lowest rates (0–24 per 1,000 live births) are found in NW Europe, North America, Australia and parts of South America.

2 How to 'compare' and 'contrast'

To compare or contrast, you need to write about *differences*, for example between areas on a map. As with making a description, you should refer to places and data wherever possible.

In comparing infant mortality between North America and South America you might write:

In North America the infant mortality is the same (0–24 per 1,000 live births) whereas in South America it varies greatly between 0–24 and 75–99 per 1,000 live births.

'How to ...'

3 How to 'explain'

To explain patterns you need to try to think of reasons why they exist. This is much more difficult and will test your understanding of geography. You may need to refer to other maps and diagrams to help you.

In explaining the high rates of infant mortality in much of Africa, shown on Figure 20, you might write:

Africa has very high rates of infant mortality because there are few doctors and hospitals and not many children are able be vaccinated. Many families live in poor conditions with little food, a lack of safe water and poor sanitation, which means that very young children find it hard to survive.

4 How to 'analyse' and 'synthesise'

An *analysis* is very similar to an explanation except that it usually involves more detail and a much greater use of facts and figures.

A *synthesis* involves the 'pulling together' of a variety of different forms of information. For example, in addition to Figure 20, you might have available maps of hospitals, doctor numbers, and access to safe water. You would be expected to look closely at all the maps and make conclusions suggesting how they link together.

'How to …'

A *A 1:50,000 OS map extract: Maidenhead*

© Crown Copyright

Mapskills Units

1 Introduction to using maps and photographs

The Millennium Stadium, Cardiff

The Millennium Stadium in Cardiff was completed in October 1999 at a cost of £130 million. It is one of the most spectacular sporting arenas in the world.

The Millennium Stadium has been used for many events since it was opened. It has been home to the FA Cup Final since the closure of Wembley Stadium, and it has also hosted rugby matches and music concerts. It has a capacity of over 70,000 people.

One of the most amazing features of the new stadium is its roof, which can be closed if the weather is poor (Figure 1).

Where is Cardiff in the UK?

Figure 2 is a location map produced by the Millennium Stadium to show the location of Cardiff in the UK. Notice that it is in South Wales close to the M4 motorway.

1 *Inside the Millennium Stadium*

make use of maps that show much more detail than is found in Figure 2.

Figure 3 on page 20 is an Ordnance Survey 1:50,000 map of Cardiff. You can see the M4 motorway to the north of the city, shown as a thick blue line. The Millennium Stadium is in grid square 1876.

Travelling from Bristol in the east, I would need to turn off the M4 at junction 30 and travel south along the A4232. Follow this route on Figure 3.

ACTIVITIES

1 Study Figure 2.
   (a) Does it take more than 1 hour or less than 1 hour to travel from Bristol to Cardiff?
   (b) Does it take more than 1 hour or less than 1 hour to travel from Carmarthen to Cardiff?
   (c) Along which motorway would you travel between London and Cardiff?
   (d) I am travelling from Birmingham to Cardiff to see a football match. I am heading south and have just passed the junction with the M42. I hear on the radio that the M50 is blocked by an accident. Which alternative motorway should I take towards Cardiff?

Where in Cardiff is the Millennium Stadium?

Let's continue my imaginary journey from Birmingham to Cardiff. How would I find the Millennium Stadium? To locate the Millennium Stadium in Cardiff, I would need to

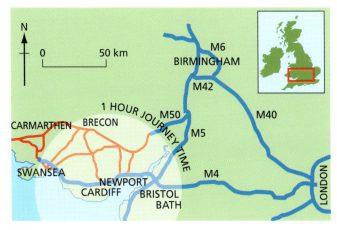

2 *Cardiff in the UK*

Mapskills Units

ACTIVITIES

2 Study Figure 3.

(a) After travelling south for just over 1 km on the A4232, I join another road. What is the number of this road?

(b) The new road is shown as a thick green line. What does this mean? (See page 123 for the symbols used on this map.)

(c) From this junction there are a number of different roads that I could follow to take me into central Cardiff to the Millennium Stadium. Look carefully at the map and find a good route. Now write down instructions for me, using details from the map (e.g. north/south, left/right, 'A' road/ 'B' road, etc.) to help me find the stadium.

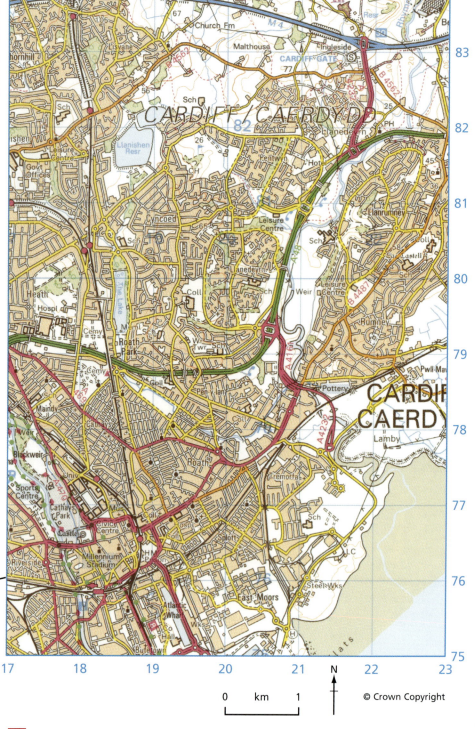

Millennium Stadium

3 1:50,000 OS map extract: Cardiff

1 Introduction to using maps and photographs

Getting closer

Whilst Figure 3 is helpful in guiding me from the M4 towards the Millennium Stadium, it doesn't help me very much as I get closer to the stadium. Is it possible to see exactly where on the map is the Millennium Stadium?

To help me further, I need to use a map with much more detail at a local scale. Figure 4 is an Ordnance Survey 1:25,000 map. Notice that this shows much more local detail than the 1:50,000 map.

Look back at Figure 3 and notice that the words 'Millennium Stadium' are written on the map. Now use the nearby roads, railway lines and stations to help you locate this same area on Figure 4. You will discover that the words 'Millennium Stadium' do not appear – not very helpful! However, if you look closely you can see a rectangular shape next to the river at 180761 which looks very much as if it could be a sports stadium.

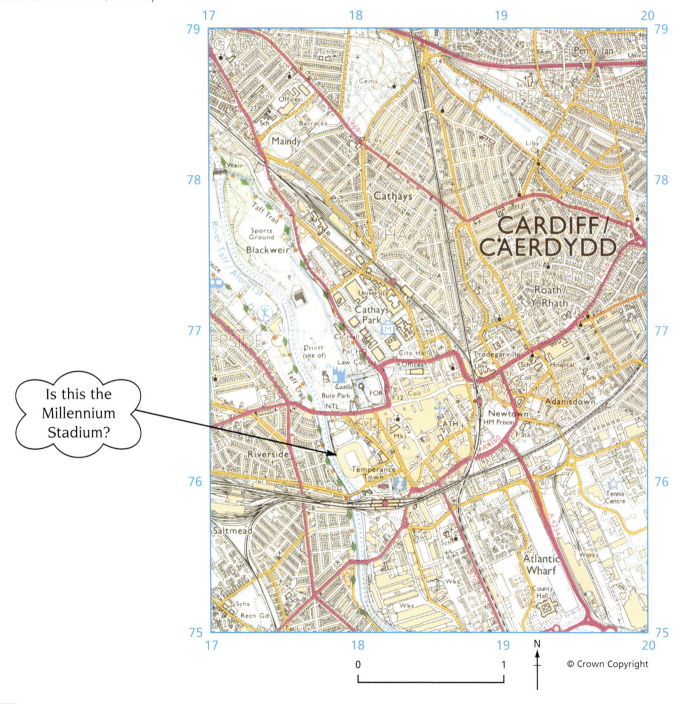

4 *1:25,000 OS map extract: central Cardiff*

Mapskills Units

5 *Oblique aerial view of Cardiff*

1 Introduction to using maps and photographs

Take to the air

To help further, we can look at an oblique aerial photograph of the area (Figure 5). It was taken from Cathays Park at approximately 183770 – find this on Figure 4. The photograph is looking south/south-west.

In the middle of the photograph you can see the remains of the castle and behind it you can see the Millennium Stadium on the banks of the river. Find these features on the map (Figure 4) to help you locate the view. The photograph helps to confirm that the rectangular shape next to the river at 180761 is indeed the Millennium Stadium.

ACTIVITIES

3 Study Figures 4 and 5.
   (a) Look at the Millennium Stadium in the photograph. Is the roof open or closed?
   (b) Look at the trees around the castle. Do you think the photograph was taken in the winter or the summer? Explain your answer.
   (c) On the photograph there is a wide bridge over the river just behind the Millennium Stadium. Use the map to suggest what the bridge is carrying.
   (d) Notice on the photograph that there is a main road running along two sides of the castle wall. Look on the map to discover the number of this road.
   (e) Use the map to find out what the square-shaped building in the bottom centre of the photograph is used for.

Almost there

Having parked my car in the centre of Cardiff I have walked to the main railway station at 183758. I now need to look at a street plan (Figure 6) to help me reach the stadium itself. My ticket tells me that my entrance into the stadium is in Park Street.

ACTIVITIES

4 Study Figure 6.
   Describe the route that I should take from the railway station to Park Street. Refer to road names – and try not to get me lost!

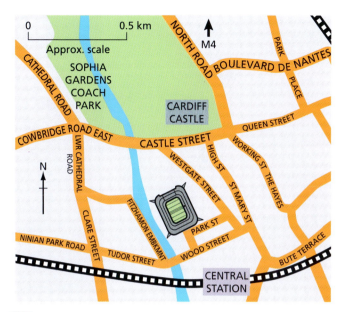

6 Street plan of the area around the Millennium Stadium

Here at last

I have now arrived at the Millennium Stadium. Figure 7 is a photograph that shows what the stadium looks like from the ground. Isn't it an impressive sight? Let's hope that my team wins!

7 The Millennium Stadium

23

Mapskills Units

2 Road map

Northern England

Three very different-sized settlements in northern England are studied in Units 3, 4 and 6.

The largest is the city of Manchester. Manchester is located some 50 km inland from the west coast port of Liverpool (Figure 1). The city of Manchester has a population of 431,000 people and it is one of the biggest settlements in the UK. Manchester grew up as a major centre for industry, particularly textiles, during the Industrial Revolution in the 19th century. Today it is a thriving modern city with several universities, an international airport and two famous football clubs. It was the host city for the highly successful 2002 Commonwealth Games.

The other two settlements that we will be studying are the town of Buxton (Unit 4) and the village of Castleton (Unit 6). Both are situated to the south-east of Manchester and to the south-west of Sheffield.

1 Satellite image of the Liverpool/Manchester area. This is a 'true-colour' satellite image. The urban areas appear light grey. Notice how Liverpool has sprawled outwards to engulf several nearby towns.

ACTIVITIES

1 Study Figure 2. We are going to take a trip from Manchester to Buxton and Castleton. Follow the route and complete the gaps, listing the answers as **(a)**, **(b)**, **(c)**, etc. in your book.

Leave the centre of Manchester and head in a south-easterly direction along the A6 towards Stockport. Just before you enter Stockport go over the M_**(a)**_. Continue on the A6 until you reach the town of _**(b)**_ Bridge. Now turn right onto the A_**(c)**_. You are now about _**(d)**_ km away from Buxton. As you travel along this road, notice the pretty _**(e)**_ Valley to your right. Soon you are in the centre of Buxton. It is very busy because _**(f)**_ (a number) 'A' roads meet here. Turn left onto the A6 again, this time heading north towards Chapel-en-le-_**(g)**_. At the roundabout take the right turn onto the A623. After a couple of kilometres at the small hamlet of Sparrowpit turn left onto the B_**(h)**_. You will soon reach the village of Castleton, which is about _**(i)**_ km from Sparrowpit.

2 Study Figure 2.
(a) What is the name of the highest hill on the map extract and what is its height in metres above sea level?
(b) The Pennine Hills in this area are in a National Park. What is its name?
(c) The National Park is the most visited of all National Parks in the UK. Make a list of some of the attractions (shown in red) to tourists.
(d) Suggest reasons why there are several large reservoirs high up in the Pennines.

2 Road maps

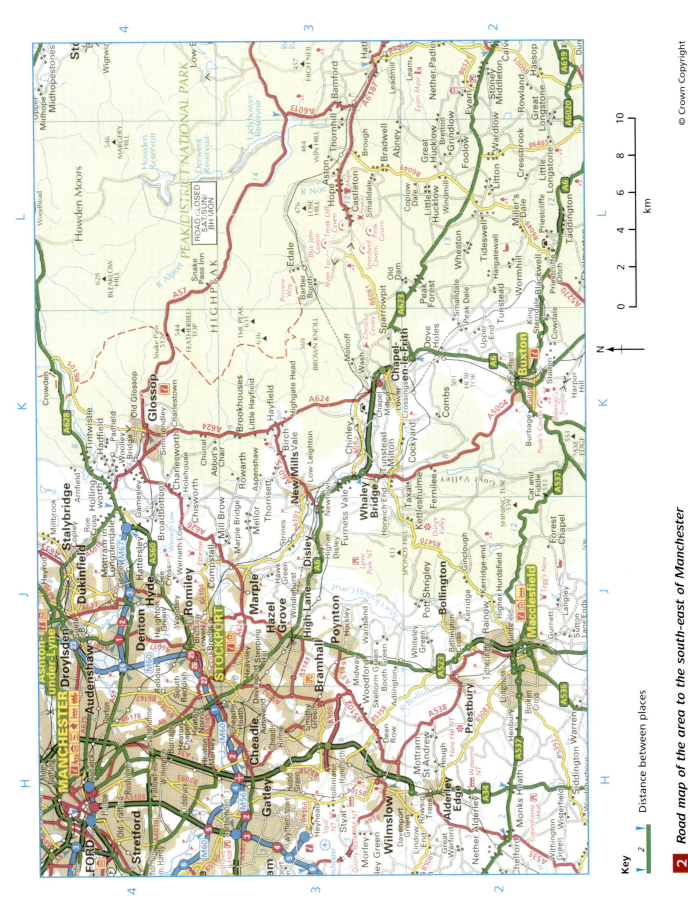

2 Road map of the area to the south-east of Manchester

Mapskills Units

3 City

Hulme, Manchester

Manchester is Britain's third largest city. It is located in north-west England to the east of the port of Liverpool (see Figure 1 on page 24). Manchester's growth was largely due to the expansion of industry, particularly involving textiles. New factories sprang up across the city and people moved in from the surrounding countryside in search of a higher income. One district that witnessed rapid industrial growth was Hulme, to the south of the city centre (Figure 1).

Hulme grew rapidly in the mid-19th century to house workers in local factories. The houses were small and built very close together. There were no gardens or bathrooms, and if toilets existed at all, they were outside the houses. Soon the area became a slum and diseases such as cholera were widespread. In the 1930s large areas of slum housing were cleared. Despite the clearances, some areas remained as thriving small communities, with shops and businesses (Figure 2).

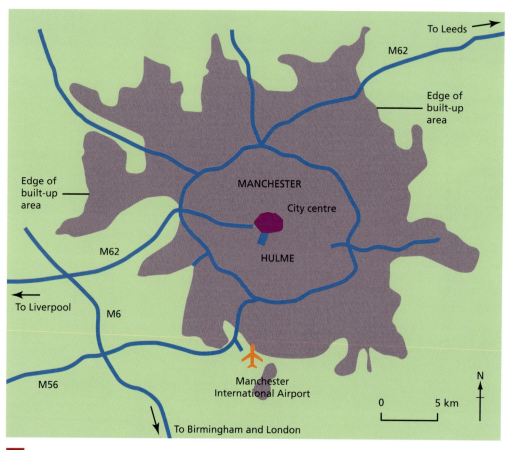

1 *Hulme in Manchester*

3 City

1960s

In the 1960s it was decided to redevelop Hulme. The flagships of the new development were four huge high-rise housing blocks in the shape of crescents (Figure 3). The idea was to mirror the lovely Georgian crescents in towns like Bath. However, their rapid and cheap construction, poor heating and lack of private space resulted in depression and ill-health amongst the residents. The flats became run-down, crime and vandalism increased, and it soon became clear that the redevelopment had failed.

1990s

The blocks were demolished and in the 1990s the area was redeveloped for the second time. However, this time the local community had a greater involvement in the plans. New houses were built to a much higher standard, each with its own private space, such as a garden or courtyard. Old buildings were restored rather than demolished, new parks and playgrounds were created, and traffic-calming schemes introduced to make the area safer for children. Offices were built (Figure 5) and new businesses developed in the area. Today Hulme is a much happier and healthier place to live.

ACTIVITIES

1 Study Figure 2.
   (a) What is the main use of the ground floor of the buildings in the photograph?
   (b) What do the many chimneys tell you about how the houses were heated in the 1950s?
   (c) Why do you think the street is not very busy?
   (d) Does the area seem to be run-down and unpleasant? Explain your answer.
   (e) Why do you think local people might have objected to the area being bulldozed and replaced with 'modern' housing?

2 Study Figures 3 and 4. The road/path running through the middle of the flats from top left to bottom right of Figure 3 is Stretford Road (see Figure 4). The approximate position where the photograph was taken is shown on Figure 4 and it is looking south-east. At the centre of the flats is the Zion Centre, which is labelled on both Figure 3 and Figure 4.

   (a) Why do you think the new blocks of flats were called 'The Crescents'?
   (b) How many 'crescents' were built?
   (c) How many storeys did each block have?
   (d) What is found at R on Figure 3?
   (e) What is the name of the road at S on Figure 3?
   (f) What is the name of the road at T on Figure 3?
   (g) The Zion Centre and a few other older buildings were not demolished to make way for the new flats. Do you think it is a good idea to keep some older buildings when an area is redeveloped? Explain your answer.
   (h) Would you like to have lived in 'The Crescents'? Explain your answer.

Weblinks

Manchester Online at www.manchesteronline.co.uk has several articles about Hulme. You will need to conduct a search of the site using the word 'Hulme'.

You'll find photographs, information and opinion from Manchester on the website 'Eyewitness in Manchester', by photographer-writer Aidan O'Rourke, at www.manchesteronline.co.uk/ewm.

Mapskills Units

2 *Preston Street, Hulme, in the 1950s*

3 City

3 *Oblique aerial view of 'The Crescents', Hulme*

Mapskills Units

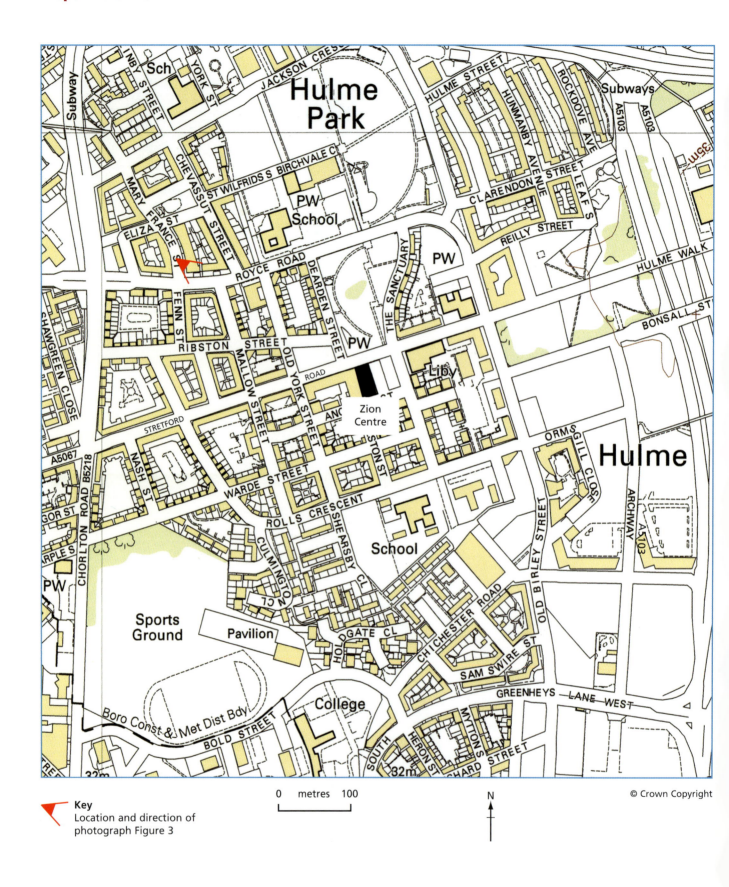

Key
Location and direction of photograph Figure 3

4 *1:5,000 OS map of Hulme, 2002*

3 City

5 *Hulme in 2003: from Stretford Road looking towards the Hulme Arch*

3 Study Figure 4.
 (a) The new Hulme has many open spaces. What is the name of the main area of parkland to the north of the map?
 (b) Why do you think planners like to create open green spaces in city centres?
 (c) Apart from the park, how else is open space used in Hulme?
 (d) In the south-eastern corner of the map extract is a recently built Asda superstore. Suggest reasons why this is a good site for a new shopping centre.
 (e) Look back to Figure 3 and locate the area of 'The Crescents' on the modern map in Figure 4. Describe the changes that have taken place since 'The Crescents' were demolished.

4 Study Figures 4 and 5. The photograph Figure 5 is taken at the junction of Royce Road and Stretford Road looking north-east to Hulme Walk (see Figure 4). The new single-span bridge is called the 'Hulme Arch'.
 (a) What is the name of the main road crossed by the Hulme Arch?
 (b) The Hulme Arch is a well-known landmark in Manchester. Why do many people think it helps to give the area a positive feel?
 (c) One of the shops in the local shopping parade in Figure 5 is a post office. Why is a post office often found in a shopping parade?
 (d) What other features in the photograph help to create a modern image of the new Hulme?

Mapskills Units

4 Town

Buxton, Derbyshire

The town of Buxton in Derbyshire is located about 35 km to the south-east of Manchester and 35 km to the south-west of Sheffield. Whilst it has a resident population of 21,000, it is visited by many thousands of tourists, especially during the summer.

What makes Buxton such a popular centre for tourism is that it is a very old town with many interesting buildings and attractions. It is known as a 'spa town'. It was first developed by the Romans who made use of the natural mineral water for bathing and for helping to cure illnesses. During the 18th century, the town became a fashionable and wealthy resort and many of its grand buildings were constructed, including the famous Opera House. One of Buxton's most famous roads is The Crescent, named after its crescent-shaped buildings. Buxton Baths (Figure 2) is in part of The Crescent.

Today, in addition to its historical features, the town offers all the usual functions and services associated with a town. There are shopping streets, a town hall, a hospital, several schools and a campus of Derby University. There is a cricket ground, a railway station and a wide variety of housing areas.

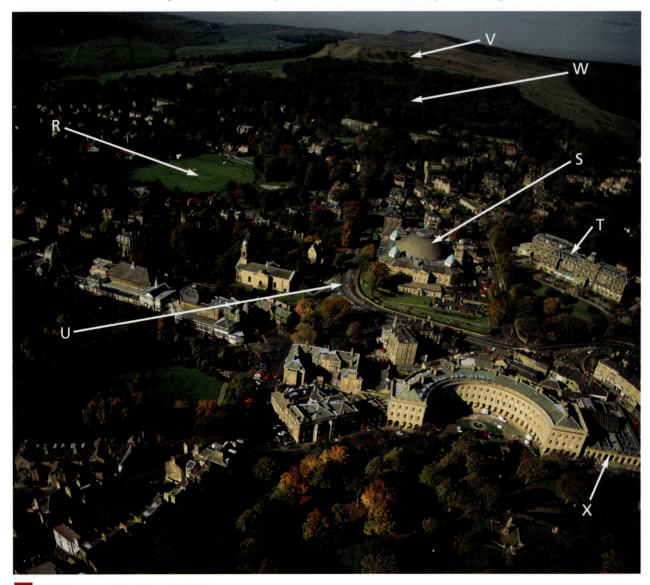

1 *Oblique aerial view of Buxton*

4 Town

ACTIVITIES

1. Figure 1 is an oblique aerial photograph of the centre of Buxton. The Crescent (with its crescent-shaped buildings) is clearly shown on both the map (near the middle of Figure 3) and the photograph. The map in Figure 3 is very detailed. It is drawn to a scale of 1:10,000.
   (a) Use a ruler to measure the distance between two horizontal grid lines on Figure 3. This distance represents 1 km on the ground. How many centimetres equal 1 km?
   (b) In what direction is the photograph in Figure 1 looking?
   (c) At what time of year do you think the photograph was taken? Explain your answer.
   (d) What is the name of the grassy area labelled R on Figure 1?
   (e) What sport is played on the grassy area?
   (f) What is the building at S?
   (g) What is the building at T?
   (h) What is the number of the road at U? Where do you think it is going to?
   (i) What is the name of the hill at V?
   (j) What is the name of the wood at W?

2. Study Figure 2. The photograph is taken looking towards the building at X on Figure 1.
   (a) Look at Figure 2. What is the name of the main building shown in the photograph? (Hint: look for a nameplate at the top.)
   (b) What evidence is there in Figure 2 to suggest that Buxton is an old town?

3. Study Figure 3.
   (a) Find out about the functions and services available in Buxton. To do this you should draw up a table with two columns. In one column you should list those functions and services aimed at local people and, in the other, those aimed at tourists.
   (b) Locate the river that flows through the centre of Buxton. Look carefully along its course to find arrows showing its direction of flow. Is it flowing from west to east or from east to west?
   (c) Locate The Park. You live in the house in Park Road directly opposite the turn for The Glade. You school is located in the bottom left of the map. The entrance to your school is on College Road. Describe the route that you take cycling to school.

2 *The Crescent, Buxton*

4. Study Figure 3.
   There are many different types of housing in Buxton. Compare the housing on the north side of Temple Road (A) with that in Dovedale Close (B). Describe the differences in the size and spacing of the houses. Compare the size of gardens. Draw simple sketches to support your answers – don't forget to draw them to the same scale!

33

Mapskills Units

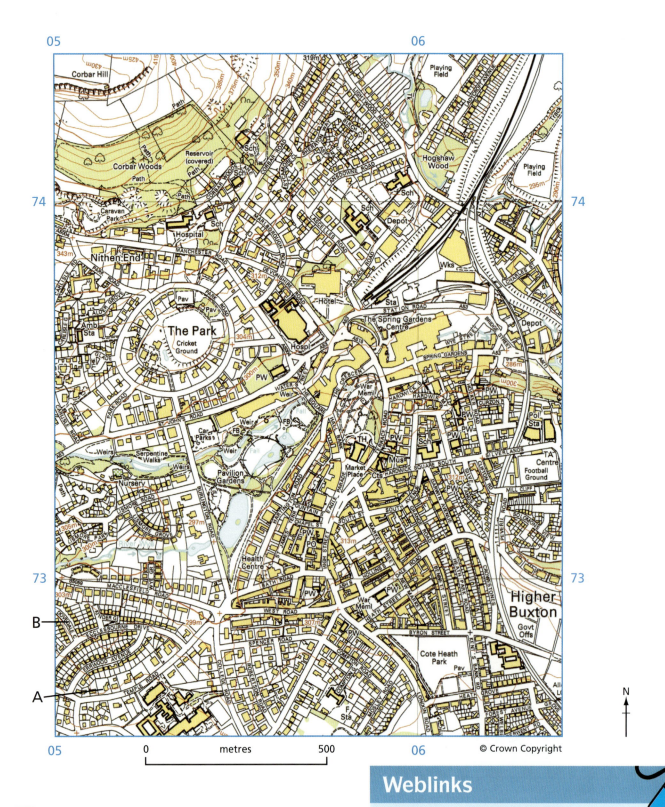

3 *1:10,000 OS map of Buxton, Derbyshire*

Weblinks

There are several websites with information about Buxton. One good site worth accessing is Buxton Online at www.buxtononline.net. Click Directory to find the main functions and services in the town.

Mapskills Units

5 Land use change

Shoreham-by-Sea, Sussex

The land use of Britain has undergone considerable changes in recent decades. In the 1930s Professor L. Dudley Stamp supervised the first Land Utilisation Survey of Britain, producing maps to a scale of 1 inch to the mile (1:63,360). It was an extremely thorough survey and was the first of its kind. In the 1960s, Professor Alice Coleman directed the Second Land Utilisation Survey, this time producing maps of England and Wales to a scale of 1:25,000.

The Land Utilisation maps have been very useful to planners and to geographers interested in studying changes in land use over time. In 1996 the Geographical Association conducted its own land use survey, Land Use-UK. This involved schoolchildren mapping their own local area. A more detailed map was produced for the Brighton/Shoreham area of Sussex to provide a direct comparison with Alice Coleman's map of the 1960s.

The map extracts in this unit show changes in land use between the 1960s (Figure 1) and 1996 (Figure 2) at Shoreham-by-Sea, just to the west of Brighton. You will need to use the two keys carefully (see pages 125–126), for although they are similar, they are not identical. Figure 3 is a recent aerial photograph of Shoreham-by-Sea.

ACTIVITIES

Take time to orientate the two maps to help you to make comparisons. Use the line of the main road running west–east through the northern edge of Shoreham in Figure 1 to help you.

1 Study both maps (Figures 1 and 2). One of the major changes to have taken place is the construction of a dual-carriageway by-passing the north of Shoreham.
   (a) What is the road number of the dual-carriageway on Figure 2?
   (b) Why do you think this new road was built?
   (c) What is the six-figure reference of the centre of the new bridge over the River Adur?
   (d) Suggest reasons why this site was chosen for the new bridge.
   (e) Much of the road has been constructed in a deep cutting. Why do you think this was done?
   (f) Locate Southwick Hill at 239075. This is an environmentally sensitive area. What have the road engineers done here to reduce any harm to the environment?
   (g) What do you notice about the line of the by-pass in relation to the built-up area of Shoreham?

2 Study Figure 2. Locate the junction of the A27(T) and the A270 in grid square 2306. Notice that a large shopping centre has been built here.

   (a) Why do you think this site was chosen for a large shopping centre?
   (b) Draw a sketch map of this grid square, adding labels to support your answer to part (a).
   (c) Imagine that planners are keen to build a leisure park, with ten-pin bowling and a cinema. Identify with a grid reference a suitable site on Figure 2 and give reasons for your choice.

3 Study Figures 1 and 2. Look at the rural land uses to the north of the built-up area. Several changes have taken place between the 1960s and 1996.
   (a) What do you notice about the sizes of the fields?
   (b) Suggest reasons for your answer to (a).
   (c) Use the keys on pages 125–126 to make a list of some of the rural land uses that were present in the 1960s but were absent in 1996.
   (d) Are there any land uses that are new to the area in 1996?
   (e) Draw two large squares to represent grid square 2207 for the two different dates. Use a pencil to mark the field boundaries in your squares. Locate Mossy Bottom Barn. Now use a colour key (ideally similar to that used on the real maps) to show the land uses. Write a few sentences outlining the changes that have taken place in this grid square between the 1960s and 1996.

35

Mapskills Units

1 *Shoreham-by-Sea 1:25,000 land use map, 1960s*

5 Land use change

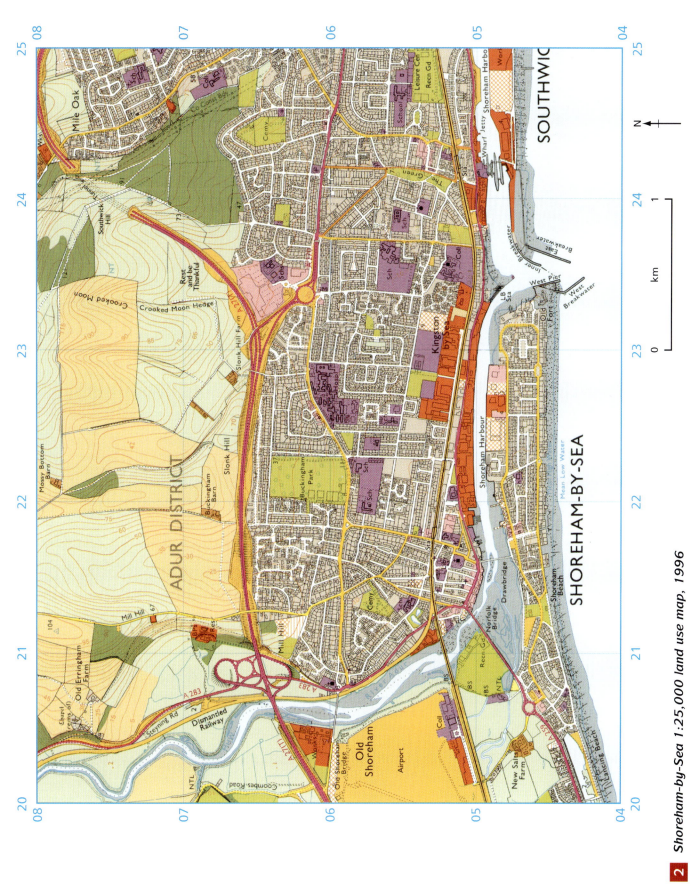

Shoreham-by-Sea 1:25,000 land use map, 1996

Mapskills Units

4 Study Figures 1 and 2. Industry is shown by the red colour on both maps.

   (a) Much of the industry is located alongside the river and in the harbour area. Why do you think this is a good location for industry?

   (b) Look at Figure 2. Name the other location where there is a large industrial works. Why is this a good location for an industry?

5 Study Figures 2 and 3.

   (a) In what direction is the photograph looking?

   (b) Give the six-figure reference of the centre of the roundabout at A.

   (c) What is the main land use at B?

   (d) Are the buildings at C New Salts Farm?

   (e) What is the land used for at D?

   (f) Which of the three bridges (E, F or G) carries the railway line over the river?

   (g) What is the name of bridge F?

   (h) What evidence is there on the photograph that Shoreham is a popular sailing resort?

   (i) Locate the housing estate at H on the photograph. Find this area on the two maps (Figures 1 and 2) and suggest when it was built.

   (j) Why do you think planners chose to build a new housing estate at H?

6 Study Figure 3.

   (a) What are the coastal defence features on the coast at J?

   (b) What is their purpose?

   (c) Do they appear to have been successful? Explain your answer.

Weblinks

An excellent review of the Shoreham land use maps, which appeared in the Ordnance Survey's *Mapping News*, can be accessed on the Ordnance Survey's website at www.ordsvy.gov.uk/education/schools/mapping_news/ winter_00_to_01/landuse.html. The maps themselves, together with Teachers' Notes, can be purchased from Latitude, 27–28 The Service Road, Potters Bar, Herts EN6 1QA (enquiries@latitude.demon.co.uk).

Maps of different scales for all parts of the country including Shoreham can be found at www.multimap.com.

5 Land use change

3 *Aerial photograph of Shoreham-by-Sea*

39

Mapskills Units

6 Village

Castleton, Derbyshire

Castleton is a village in the Peak District, about 35 km south-east of Manchester and 18 km west of Sheffield (see Figure 2 page 25). It has a resident population of about 700 people. Castleton is situated at the foot of a limestone escarpment close to where natural springs emerge from the rock. Originally the escarpment would have provided shelter, and the springs fresh drinking water, for the settlers. These were two important factors affecting the choice of site for the village of Castleton.

Figure 3 is a very detailed map of Castleton. It is at a scale of 1:5,000. Notice that several roads meet in the centre of Castleton. This helps to explain why the village is very compact. Such a pattern is known as 'nucleated' and it is very common in the UK.

Castleton has a long and interesting history. As its name suggests, there is a nearby castle, which you can see in Figure 1. There are also some ancient defence works in the form of a ditch, part of which can still be traced at the edge of the village. More recently, several industries grew up in the village. The river provided power for a corn mill and there was also a cotton mill and a saw mill. Rope and candle makers supported the nearby lead mines. A mineral locally known as 'Blue John' is made into jewellery and other ornaments by craftsmen in the village. Nowadays there is a large limestone quarry about 1 km to the south-east of Castleton, which provides employment for people in the village.

With its varied history, Castleton has become a popular tourist destination, particularly for day-trippers. Each year over 2 million people visit Castleton and the nearby Hope Valley. There are many opportunities for walking and visiting nearby caves, and Castleton itself has several gift shops, cafés and restaurants catering for the needs of visitors. Whilst tourists bring money and employment, they also bring problems in the form of traffic congestion, pollution and footpath erosion.

ACTIVITIES

1 Study Figures 1 and 3. The aerial photograph in Figure 1 is taken from the south-west corner of the map extract (Figure 3) looking towards Castleton.

(a) In what direction is the photograph looking?

(b) What is the name of the ruined castle at R on Figure 1?

(c) Why do you think the castle was sited at this point?

(d) What is the name of the steep-sided valley in the photograph? (See Figure 8 on page 10 to remind you of the contour pattern that shows a valley.)

(e) Find the village church labelled S on Figures 1 and 3. What symbol is used on the map to show this building?

(f) Locate V on Figures 1 and 3. What is the land here used for?

(g) Why do you think the land at V has not had houses built on it?

(h) What is the name of the road labelled T?

(i) What type of farming do you think the land around the village is used for? Explain your answer.

(j) What is the name of the road at W?

6 Village

1 *Aerial view of Castleton, Derbyshire*

2 Study Figures 2 and 3. The photograph was taken from point X on the map extract (Figure 3) looking north.
  (a) At what time of year do you think the photograph was taken? Explain your answer.
  (b) What is the name of the road on which the photographer was standing to take the picture?
  (c) Use the map extract (Figure 3) to find the name of the street to the far right of the photograph.
  (d) Do you think the cottages in the photograph are old or new? Give reasons for your answer.
  (e) Do you think they were built for relatively rich or poor people? Explain your answer.

41

Mapskills Units

3 Study Figure 3.
   (a) What is the name of the ancient ditch that marks part of the edge of Castleton?
   (b) Give two street names which suggest that Castleton had an industrial past.
   (c) If you walked along Back Street from north to south, would you be going uphill or downhill? What map evidence did you use to answer this question?
   (d) What is the number of the 'A' road that runs through Castleton?
   (e) What is the name of the river that runs through the village?
   (f) We have already seen that Castleton has a church. Use the map to make a list of any other services and functions available in the village.

Weblinks

The Peak District National Park maintains a superb website (www.peakdistrict-education.gov.uk) with lots of information about Castleton, including population data, information about tourism and a list of goods and services available in the village, at www.peakdistrict-education.gov.uk/Fact%20Sheets/fz7cas.htm. Maps at a scale of 1:5,000 (Figure 3) can be site-centred for anywhere in the UK. For further details contact Latitude at 27–28 The Service Road, Potters Bar, Herts EN6 1QA (enquiries@latitudemaps.demon.co.uk).

2 *Castleton War Memorial, Market Place*

6 Village

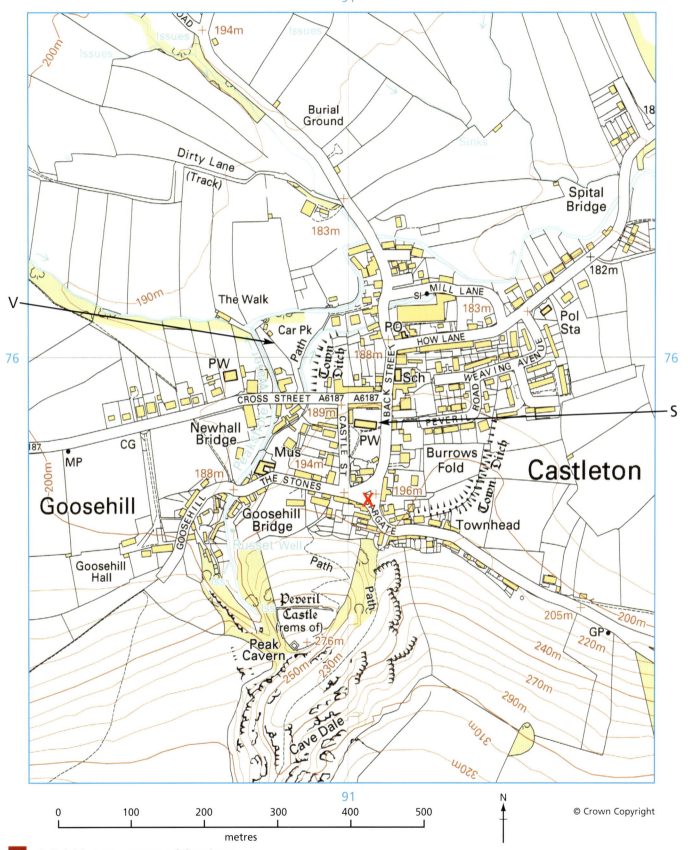

3 1:5,000 map extract of Castleton

7 Limestone scenery

Ingleborough, North Yorkshire

Limestone is one of the most common rock types in the UK. It is a sedimentary rock and is very rich in the mineral calcite. There are several different types of limestone. One of the most widespread is Carboniferous limestone, so-called because it was formed during the Carboniferous geological period some 300 million years ago.

Carboniferous limestone is a tough grey rock that forms much of the Pennine Hills in central and northern England. Some of the typical landscape features produced by Carboniferous limestone are shown in Figure 1.

One important characteristic of limestone is that it is permeable. This means that it allows water to pass through it along the vertical joints and horizontal bedding planes (the junctions between the separate layers of limestone). This means that there are few rivers on the surface of limestone.

Another characteristic of limestone is that it reacts with acidic rainwater and slowly dissolves. As water flows through the limestone, the cracks are widened until huge underground caverns are formed (Figure 5). When the water drips from the roof, the dissolved calcite precipitates out of solution to form features that look like icicles. These are stalactites. If the water drips onto the cavern floor, stumpy stalagmites are formed. You can see some of these features in Figure 6.

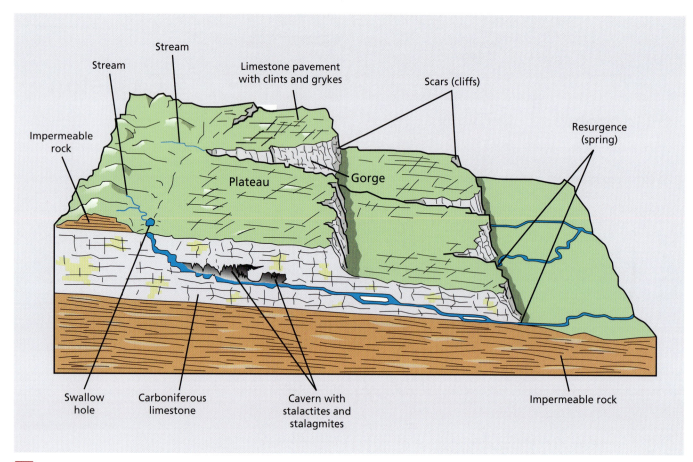

1 *Typical features of limestone landscapes*

7 Limestone scenery

ACTIVITIES

1 Study Figure 2. The photograph was taken from the rocky outcrops to the north of Ingleborough looking south towards it (see Figure 3).
   (a) Describe the shape of Ingleborough hill. Draw a simple sketch to show the outline of the hill.
   (b) The bare rock surface in the foreground is called a limestone pavement (see Figure 1).
     • Does the upper surface look smooth or rough?
     • Are the edges generally smooth or angular?
     • Is it generally flat or sloping?
     • Is there any pattern to the cracks?
     • Where is grass growing? Can you suggest why it is here?
   (c) Can you see any signs of human activity in the photograph? Suggest reasons for your answer.
   (d) Make a list of five adjectives (describing words) that describe what the landscape looks like to you.

2 *Limestone pavement with Ingleborough in the background*

Mapskills Units

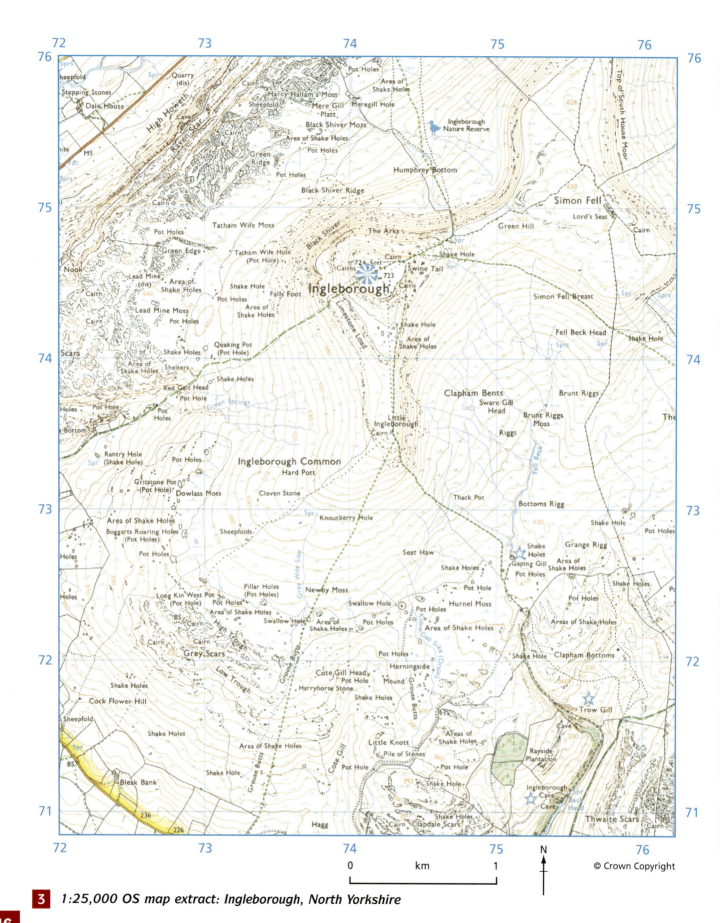

3 *1:25,000 OS map extract: Ingleborough, North Yorkshire*

7 Limestone scenery

2 Study Figure 3. You are going to follow a route up and over Ingleborough to see some of the typical features of Carboniferous limestone. You start your walk at 720733 and head along the bridleway (green broken line) in a north-easterly direction towards Ingleborough. Eventually, after walking steeply uphill, you reach the top of Ingleborough hill.

(a) What is the height of the hill top?
(b) What is the meaning of the blue symbol? Why do you think it has been put at this point on the map?
(c) Give the four-figure reference for the top of Ingleborough hill.

Now continue on for about 200 m before taking the footpath heading due south towards Little Ingleborough. Stop here.

(d) At Little Ingleborough, is the steepest downhill slope to the west or to the east? Explain your answer.

Now continue along the left fork and stop at Gaping Gill (see Figure 4).

(e) What is the six-figure reference of Gaping Gill?
(f) Use Figure 1 to find the name of this typical limestone feature.
(g) What is the name of the small stream that is plunging down Gaping Gill?
(h) Many of the surface streams in this area disappear underground at about 400 m. Use Figure 1 and the text to help you explain why the streams are doing this.
(i) What do you think the people in the photograph (Figure 4) are doing?

Continue to follow the footpath in a southerly direction to pass through Trow Gill (Figure 5).

(j) Describe the valley sides in the photograph.
(k) Is the valley narrow or wide?
(l) Refer to Figure 1 and suggest a geographical name for this type of valley.

Continue along the footpath until you reach Ingleborough Cave. Notice that there is a spring here.

(m) What is a spring and how does it form?
(n) The photograph Figure 6 shows the features that are typical of a cave in limestone country. Describe the features of the cave and explain how they were formed.
(o) Work out how far you have walked in total, to the nearest kilometre (see pages 8–9 for help).
(p) If you could visit only one of the places shown in the photographs, which would it be, and why?

4 *Gaping Gill*

47

Mapskills Units

5 *Trow Gill*

> ### Weblinks
> Some further details about the Ingleborough area, especially Ingleborough Cave, can be found at the Yorkshirenet site at www.yorkshirenet.co.uk/clapham/.

6 *Underground features in a limestone cavern*

Mapskills Units

8 River flood

Malton, North Yorkshire

In 1999 and 2000 the small market town of Malton, together with the neighbouring town of Norton, was affected by serious river flooding. Roads were cut off, the railway was under water for several weeks and many homes and businesses were inundated.

In March 1999 heavy rain and melting snow combined to cause the River Derwent to burst its banks in several places, sending muddy floodwaters pouring into people's homes (Figure 1). A total of 100 homes and 40 businesses were flooded.

In November 2000 the area was once again affected by flooding (see Figure 3). There had been several weeks of heavy rainfall which had saturated the soil. Unable to cope with all the surface water, the river overflowed, flooding many of the same areas hit in the previous year.

River flooding is quite common in the UK, especially in the winter and in spring when there is often heavy rain and snowmelt. Flooding causes great damage to homes and is very upsetting to people who are affected. Whilst few people die as a result of flooding in the UK, the cost of damage caused by a flood often runs into several millions of pounds.

The Environment Agency tries to prevent flooding by building flood defences and by issuing flood warnings to the public. In Malton, the Environment Agency is building floodwalls and banks to try to prevent a repeat of the recent floods.

ACTIVITIES

1 Study Figure 1.
   (a) Write a couple of sentences describing what the photograph shows.
   (b) Describe the likely feelings of the people in the house.
   (c) What have they done to try to stop the water getting into their home?
   (d) When the water level drops again, what will be left deposited on the ground? What problems will this cause?
   (e) The water is about ankle deep. Imagine that there was water to the same depth in your house. Describe the likely damage that this would do to your family's possessions.

1 Homes flooded in Malton in March 1999

Mapskills Units

2 1:25,000 OS map extract: Malton © Crown Copyright

2 Figure 3 shows the flood of 2000 in the centre of Malton. Locate the area shown in the photograph on the map. To do this, use the location of the bus/coach station and the railway station to help you. Both are labelled on the photograph and, with the aid of the key on page 124, you should be able to find them on the map extract (Figure 2).
   (a) Draw the map symbol for a bus or coach station. What is its six-figure grid reference?
   (b) Draw the map symbol for a railway station. What is its six-figure grid reference?
   (c) Having located these two features on the photograph and on the map, try to suggest in which direction the photograph is looking.

3 Study Figures 2 and 3.
   (a) Use the map to help you locate the river on the photograph. Now look at what appears to be a straight canal towards the bottom of the photograph. Use the map to discover what this is.
   (b) What do you think is partly underwater at X?
   (c) How has the flooding affected people living in the houses at Y?
   (d) How has transport by road and rail been affected by the floods?

50

8 River flood

3 *Malton in November 2000*

Weblinks

The Environment Agency at www.environment-agency.gov.uk/ has a huge amount of information about the flood hazard (click 'Flood' in the Choose a Section box). You can enter your postcode to see if you are at risk from flooding!

The *Yorkshire Post's* website at www.ypn.co.uk may help you with Activity 4.

4 Imagine that you are a reporter for the local newspaper, the *Yorkshire Post*. You have spent the day in Malton (probably getting your feet wet!) and your editor has asked you to write a front-page story about the effects of the flood of November 2000 on the town. You are only allowed to use 100 words and must include a 'snappy' headline. Your account will go alongside a large copy of Figure 3 on the front page of the newspaper. You may like to use an ICT package, such as Publisher, to produce your article.

51

Mapskills Units

9 Coastal erosion

The Green Bridge of Wales, Pembroke

The Pembrokeshire National Park in Dyfed, Wales contains some of the most dramatic coastal scenery in the UK. There are steep cliffs, narrow inlets and wide sandy bays.

One of the best-known landforms on the south coast is a natural arch called The Green Bridge of Wales (Figure 1). The arch is about 24 m high and has a span of more than 20 m. To the right of the arch is an isolated column of rock – this is a landform called a stack. In the past it was part of another arch, but the roof collapsed to separate it from the rest of the coast. To the left of The Green Bridge of Wales you can see caves at the base of the cliff. Figure 2 describes some common features of coastal erosion.

The rock that forms the coastal features in Figure 1 is limestone. It is a very tough sedimentary rock – you can see the layers or beds that help us identify the rock as being sedimentary – which is why it forms dramatic cliffs.

The photograph was taken at low tide. Look closely at the cliff to see a darker green/black staining of the rock, which shows how high the water rises at high tide. Many of the flat rocks and boulders are covered by water at high tide.

The sea looks very calm in the photograph but this stretch of coastline suffers powerful wave attack during storms. The waves crash against the cliffs, undercutting them and prising apart cracks and joints in the rocks. Rockfalls are common as the cliff face collapses, forming piles of boulders at the cliff foot (see Figure 1).

ACTIVITIES

1 Study Figure 1.
   (a) Look closely at the layers of rock exposed on The Green Bridge of Wales. Do they appear to be tilting (dipping) towards the sea or towards the land?
   (b) Where do you think the boulders at X have come from?
   (c) How do you think the boulders got there?
   (d) What evidence is there on the photograph to suggest that it was taken at low tide?
   (e) What type of vegetation is growing on the cliff top?
   (f) Can you suggest why there are no trees on the cliff top?

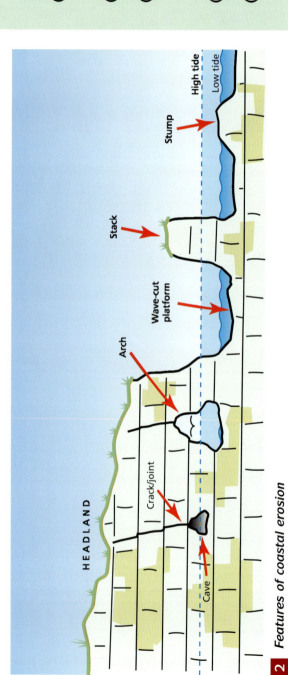

2 *Features of coastal erosion*

9 Coastal erosion

1 *The Green Bridge of Wales*

Mapskills Units

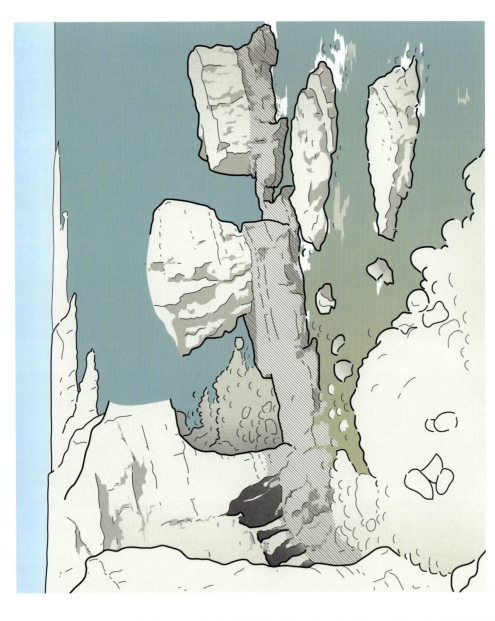

2 Figure 3 is a partly completed sketch of the main coastal features shown in the photograph (Figure 1).

(a) Make a copy of Figure 3 and carefully complete the sketch. (Look back to pages 13–14 to remind yourself how to draw a sketch from a photograph.)

(b) Now label the following features on your completed sketch:
- Steep cliffs
- Boulders at the foot of the cliffs
- Caves
- Beds of limestone
- Arch ('The Green Bridge of Wales')
- Stack
- Stump (see Figure 2)

(c) Give your sketch a title.

3 Figure 4 is a 1:25,000 map extract of the stretch of coastline that includes the features shown in Figure 1.

(a) What is the four-figure grid reference for The Green Bridge of Wales?

(b) What is the spot height of the cliff top in this square?

(c) If you were to visit The Green Bridge of Wales by car, you would probably park in the car park nearby. What is its six-figure grid reference?

(d) Just to the east of The Green Bridge of Wales is a group of stacks called Elegug Stacks. What is meant by the blue star symbol that is next to these features?

(e) Why is the area to the north of The Green Bridge of Wales a DANGER AREA?

(f) The Pembrokeshire Coast Path crosses the map extract. Draw the symbol used to show this important long-distance footpath as it appears in grid square 9394.

(g) How far would I have to walk along the Pembrokeshire Coast Path from the car park at Stack Rocks to the natural arch at Bullslaughter Bay?

(h) If I wanted to have a picnic in a sandy bay, where would you suggest that I go? Give a six-figure grid reference to help me find it.

3 *Sketch of coastal features in Figure 1*

9 Coastal erosion

4 1:25,000 OS map extract: South Pembrokeshire coast

4 Design a poster to be displayed on an information board at Stack Rocks car park. The aim of your poster is to help people understand more about The Green Bridge of Wales. Tell people what to look for, and help them to understand what has happened to form the various features. Include a sketch of The Green Bridge of Wales or a labelled photograph copied from the internet.

Weblinks

Take a look at Pembrokeshire National Park at www.pembrokeshirecoast.org.uk/.

A good photograph of The Green Bridge of Wales can be found on the West Wales Photograph Album site at www.westwales.co.uk/graphics/grenbrid.jpg.

55

Mapskills Units

10 Tourism

The Eden Project, Cornwall

At first sight the Eden Project, with its enormous plastic 'bubbles', looks like a giant theme park (Figure 1). It is, however, a wonderful collection of plants from around the world, mostly housed in vast circular greenhouses.

The Eden Project is the brainchild of one man, Tim Smit, who wanted to develop an abandoned Cornish china clay quarry into a modern exhibition centre. Through its many displays, paintings and sculptures, the Eden Project stresses the importance of plants in our everyday lives. It shows how plants are used for food, in industry, in building and in medicine.

At its heart are the two 'biomes', each made up of four interconnecting domes (see Figure 2). The term 'biome' means 'large-scale ecosystem'. One biome represents the warm temperate zones (such as the Mediterranean) and the other the humid tropics (the tropical rainforests). Together there are over 500,000 plants in the two biomes. The domes are made up of giant hexagons (Figure 2) each up to 11 m across, covered with thick plastic sheeting, which is lighter and tougher than glass. The largest dome is so huge that you could fit the Tower of London into it!

The Eden Project promotes sustainable development. This means using the world's resources in such a way as to make sure there will always be resources for future generations to enjoy. It means not wasting resources and not harming our environment.

From the time it opened in March 2001 it attracted well over 2 million people in less than 18 months (see the website to find out its current visitor total). It is now one of the UK's more important tourist destinations and it has had a huge effect on the economy of Cornwall. A third biome is now being planned, to represent the world's deserts.

ACTIVITIES

1 Study Figure 1.
   (a) Describe the view in the photograph.
   (b) Do you think it looks like a giant theme park? Explain your answer.
   (c) What evidence is there in the photograph that the Eden Project is sited in an abandoned quarry?

1 The Eden Project

10 Tourism

2 *Oblique aerial view of the Eden Project*

Mapskills Units

3 *Plan of the Eden Project*

2 Study Figures 2 and 3.
   (a) Using the compass on Figure 3 to help you, suggest which way the photograph (Figure 2) is looking.
   (b) Which of the two biomes is at the bottom of the photograph?
   (c) What name is given to the winding path at P on Figure 2 that leads down from the Visitor Centre to the main site?
   (d) Why does the path take such a winding course rather than just going in a straight line down the quarry side?
   (e) What happens in the tented area Q on Figure 2?
   (f) How are the domes designed to allow air to flow through them? (Hint: look at the top of the domes!)
   (g) Locate The Link on Figure 3. Now find this on the photograph. What is unusual about its roof?
   (h) Name three types of eating place where food can be bought in The Link.
   (i) There is a land train on the site (see Figure 3). Where does the train travel to and from?

10 Tourism

3 Study Figure 4. The Eden Project is mostly found in grid square 0454.
  (a) Give the six-figure grid reference of the Humid Tropics Biome.
  (b) Road access to the Eden Project is along local minor roads. What problems do you think might result from the huge number of people who wish to visit the Eden Project?
  (c) What evidence is there on the map that this part of Cornwall has a history of mining?

4 Study Figure 4. The Eden Project is sited in an abandoned china clay quarry. China clay is white clay that is used as a 'whitener' in making porcelain, paper and even toothpaste. There are several working quarries in the area. Part of the processing of china clay involves drying. This takes place in nearby Par (grid reference 0753).
  (a) Describe the site of the drying works.
  (b) Why do you think the works are sited here?
  (c) Find two forms of transport near the works.

Weblinks

The Eden Project's official site is at www.edenproject.com.

4 1:25,000 OS map extract: the Eden Project

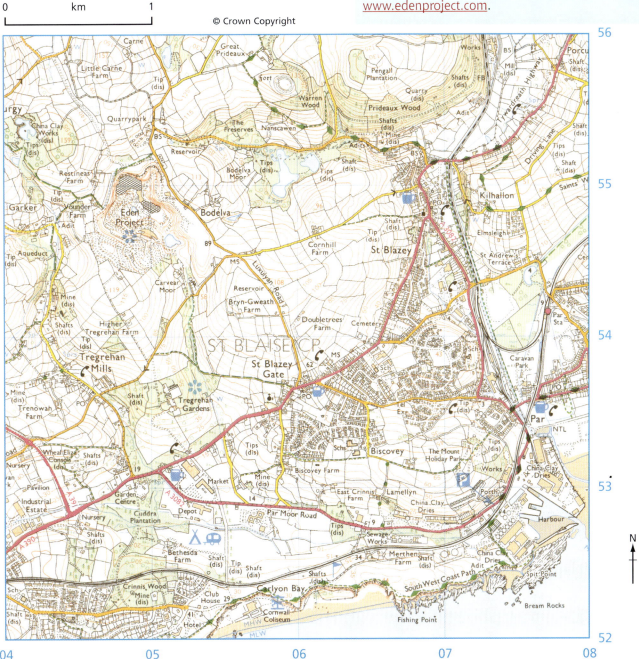

© Crown Copyright

Mapskills Units

11 Rivers (1)

The Rhône, Switzerland

The River Rhône is one of the longest rivers in Europe. It flows for some 800 km from its source high up in the Swiss Alps to its mouth in the Mediterranean near the French port of Marseilles (Figure 1). In the following three units we are going to study some of the major features of the River Rhône and its valley.

The source of the River Rhône is rather unusual. It is in fact meltwater from the Rhône Glacier (Figures 2 and 3). When the winter snow and ice melts in spring and summer, water pours from beneath the Rhône Glacier to form the headwaters of the Rhône, locally called the Rotten (see Figure 3).

Meltwater rivers carry large amounts of fine sediment, which result in their appearing 'milky' to look at. You can see this characteristic in the photograph in Figure 2. The amount of water flowing in these rivers varies considerably with the seasons. It is usually greatest during the middle of the day and during the summer. Rivers that carry lots of sediment and whose flow varies a great deal tend to split into several channels – this is called braiding.

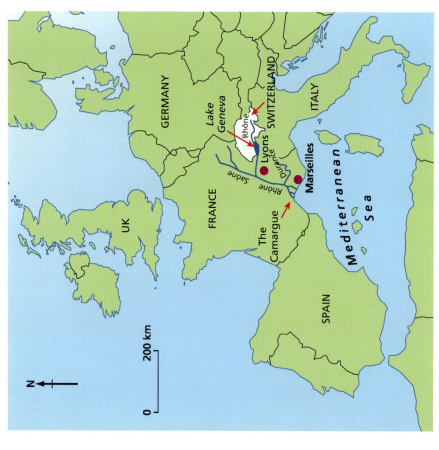

ACTIVITIES

1 Study Figure 2. Locate the Rhône Glacier about halfway up the mountainside in the centre of the photograph.

(a) Describe the steepness of the river channel as it flows out of the glacier.

(b) Attempt to estimate the width of the river towards the bottom of the photograph.

(c) Describe the colour of the river. What causes the river to be this colour?

(d) Describe the flow of the river. Is it smooth or turbulent (rough)?

(e) Describe the nature of the land on either side of the river. Is it steeply sloping, flat, hilly, etc?

1 *The course of the River Rhône*

(f) What evidence is there that the land alongside the river is used for farming?

(g) What type of vegetation can you see on the lower slopes of the mountains?

(h) What evidence is there that the area is popular with tourists?

(i) Why do you think tourists come here?

60

11 Rivers (1)

2 *Source of the River Rhône*

61

Mapskills Units

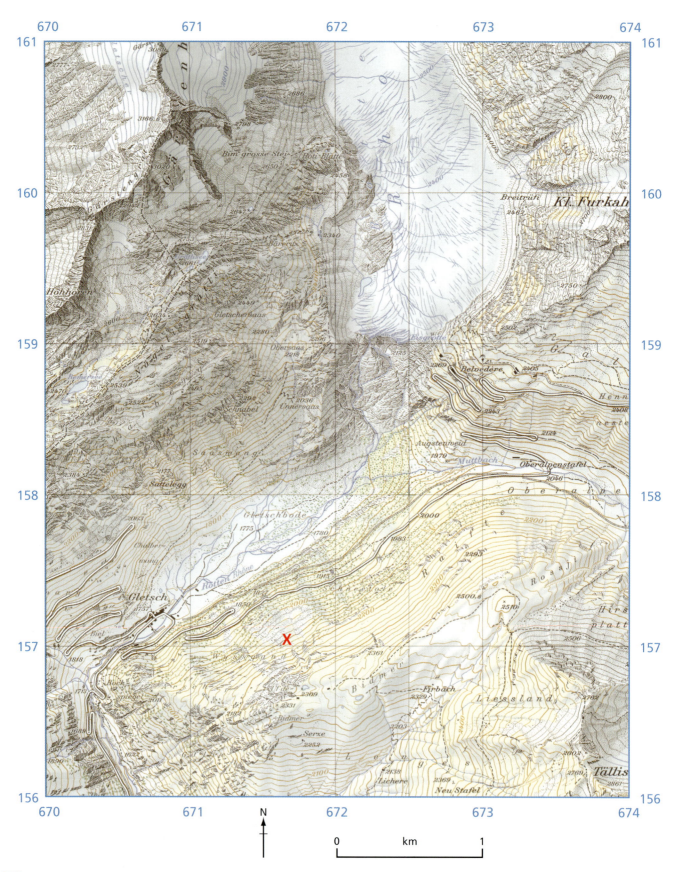

3 *1:25,000 map extract showing the source of the River Rhône*

11 Rivers (1)

2 Study Figure 3.
   (a) How is the Rhône Glacier shown on the map?
   (b) The blue lines running across the glacier are contours. The heights are given in metres above sea level. What is the interval between each contour?
   (c) What do you think the other blue lines on the glacier are?
   (d) Locate the stream coming out of the front (snout) of the glacier. What is the height of the nearby spot height?
   (e) Follow the river away from the glacier. In which direction is it flowing?
   (f) The river is soon joined by another river flowing from the east. What is the name of this river?
   (g) Use the contours to give an approximate height above sea level where the two rivers join.
   (h) Between the point where the two rivers join (the confluence) and the small village of Gletsch, the river splits into several channels. What is this called? Draw a simple sketch to show what happens here.
   (i) Describe the course of the roads around Gletsch. Why do you think they are like this?

3 Study Figures 2 and 3. The photograph in Figure 2 was taken from the bridge over the river just to the north-east of Gletsch (see Figure 3).
   (a) Which way is the photograph looking?
   (b) What symbol is used on the map to indicate the track to the left of the river?

4 Study Figures 2 and 3. Find the zig-zag road to the right of the Rhône Glacier on the photograph and locate it on the map.
   (a) What is the name given to the hotel about halfway up this zig-zag road?
   (b) Why does the road zig-zag up the mountainside?

5 Study Figure 3. Attempt to draw a cross-section across the valley of the River Rotten (Rhône). Take your section from Sattelegg (2177 m) to the north of Gletsch, across the valley to the 2100 m contour at X. Use the same scale shown in Figure 4 and follow the steps described earlier in the book on pages 10–11. Give your section a title, label the axes, the river, its valley floor and the road.

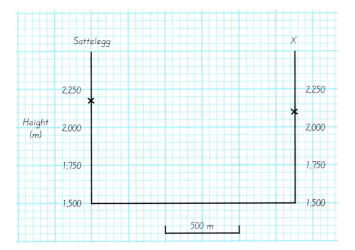

4 *Graph axes for drawing cross-section across the Rhône valley near Gletsch*

Weblinks

There are several websites that give further information about the Rhône Glacier and the small village of Gletsch. One site worth looking at is Switzerland.isyours.com at
http://switzerland.isyours.com/e/guide/valais/alpinepasses.html

– but there are many more.

63

Mapskills Units

12 Rivers (2)

The Rhône, France

As the River Rhône flows through France it shows more signs of being managed by people. Some stretches of the river have been canalised. This means that straight sections have been constructed to by-pass the main river. You can see this in Figure 1. The river has been managed in this way to help reduce flooding and also to enable dams to be built for the production of hydro-electric power.

Figure 2 is a satellite photograph of the River Rhône as it flows northwards, about 60 km to the east of the city of Lyons (see Figure 1 page 60). The top of the map extract in Figure 1 is at the bottom of the satellite photograph. Read the caption to Figure 2 to understand what the colours indicate.

The river is now meandering over a wide and flat floodplain. Much of the land on either side of the river is farmed – you can see the fields in Figure 2. In places the land is so flat that the river is forced to split into smaller channels. You can see this clearly in the centre of the photograph.

The most interesting river feature in the photograph is labelled X. It is an old meander that has been 'cut off' to form an ox-bow lake. Records indicate that it formed in 1690. Since then, it has gradually silted up and vegetation has begun to grow between its banks. It is the vegetation, growing in damp conditions, that marks it out from the cultivated fields alongside.

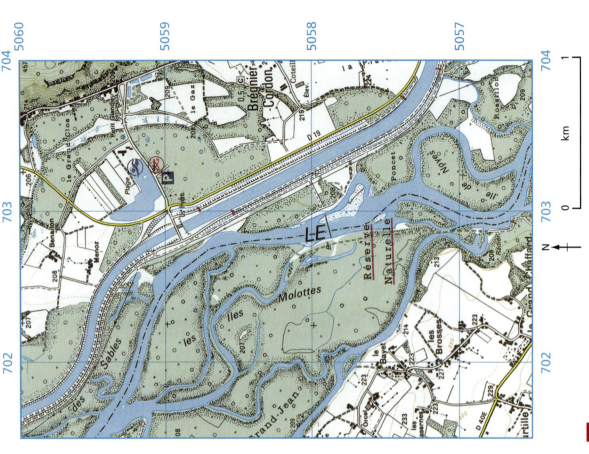

1 *1:25,000 map extract: area south of Brangues*

Activities

1 Study Figure 1.
  (a) Look closely at the straight section of the river. Notice that it has raised banks (called levées). Why do you think engineers have constructed these embankments?
  (b) Locate the dam. What evidence is there that it is used to produce electricity?
  (c) Why do you think engineers chose to construct a dam on a canalised section of the river rather than on a natural section?
  (d) What are the attractions to tourists wishing to visit the dam?
  (e) Why do you think the old course of the river has become a 'Réserve Naturelle'?

64

12 Rivers (2)

2 Satellite photograph of the River Rhône near Brangues. The river shows up as blue. Vegetation is green and pink, making the pattern of fields clearly visible.

Mapskills Units

2 Study Figures 2 and 3.
   (a) What is the name of the river feature at A?
   (b) Describe the shape of the fields at B.
   (c) Use Figure 3 to discover the name of the small tributary river at C.
   (d) What shade of green is used to show forest on the photograph?
   (e) Why do you think the ox-bow lake stands out so clearly on the photograph?

3 Study Figure 3.
   (a) Locate the ox-bow lake on the map extract. Use the key to identify the types of vegetation that are found within the banks of the old meander.
   (b) What is the name of the small settlement inside the curve of the ox-bow lake?
   (c) What is the height (to the nearest whole metre) above sea level of the settlement?
   (d) How does the shape of the village of la Sauge help to mark the edge of the ox-bow lake?
   (e) What evidence is there that the river has deposited huge amounts of sediment just to the west of the ox-bow lake?

   (f) What type of woodland is the Forêt d'Evieu?
   (g) What evidence is there in grid square 698 5065 that the River Rhône is used for transport?

4 Study Figure 3. Draw a sketch to show the main geographical features on the map extract.
   (a) Draw a frame the same shape as the extract and add the map gridlines.
   (b) Carefully draw the river, using the gridlines to guide you.
   (c) Draw the outline of the ox-bow lake.
   (d) Draw the tributary river (labelled C).
   (e) Draw the Forêt d'Evieu.
   (f) Locate the settlements of le Sauget, la Sauge, Brangues and Groslee.
   (g) Draw the local (yellow) roads and bridges over the river.
   (h) Add labels to identify the main features on your sketch (e.g. meanders, ox-bow lake).
   (i) Complete your sketch by writing a title and adding a north point and approximate scale.

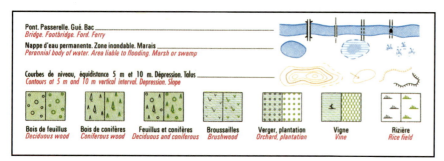

Key for map extracts

Weblinks

To find out how ox-bow lakes are formed, access the BBC's Bitesize revision online guide at www.bbc.co.uk/schools/gcsebitesize/geography/water/meandersandoxbowlakesrev1.shtml. There is an animation here that will show you how a meander becomes an ox-bow lake.

12 Rivers (2)

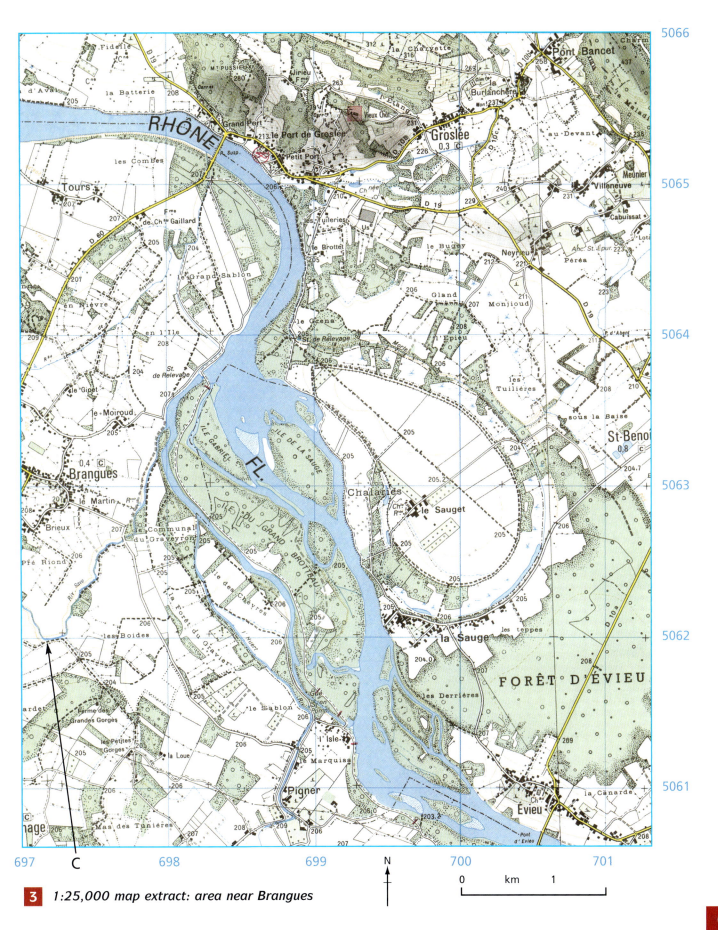

3 1:25,000 map extract: area near Brangues

Mapskills Units

13 Rivers (3)

The Rhône delta, Camargue, France

After a journey of some 800 km, the meltwater that formed the source of the River Rhône in the Swiss Alps finally reaches the Mediterranean Sea near Marseilles in France (see Figure 1 on page 60).

Over thousands of years, the mouth of the River Rhône has become choked with vast amounts of deposited sediment. This deposited river silt (called alluvium) has formed a large flat, marshy area called a delta (Figures 1 and 2). Deltas are areas of new land that extend into the sea. They are common landforms and are found across the world: you have probably heard of the Nile delta in Egypt.

The Rhône delta is called the Camargue. It is a very special area. The flat landscape, with its beautiful lakes (called *étangs*) and marshy vegetation, is home to many species of plants, birds and animals including the famous white horses (Figure 3) and majestic pink flamingos.

The Camargue is a Regional Nature Park and it is carefully managed to conserve the wildlife. However, it is a working environment too, and large parts of the area are farmed, particularly for rice. Salt is also produced in the Camargue. Lagoons are flooded with seawater, which is then allowed to evaporate, leaving behind salt deposits (Figure 4). Some people are concerned that intensive farming and industrial developments may damage the natural environment.

1 *The Rhône delta*

13 Rivers (3)

ACTIVITIES

1 Study Figures 1 and 2.
   (a) What is the name of the town at A?
   (b) Just above A the Rhône splits into two. What are the names of the two branches of the river that flow to the Mediterranean?
   (c) What is the name of the large lake at B?
   (d) What is the name of the small coastal resort at C?
   (e) What is the approximate straight-line distance between A and C?
   (f) Why do you think satellite photographs like Figure 2 are so useful in drawing maps like Figure 1?

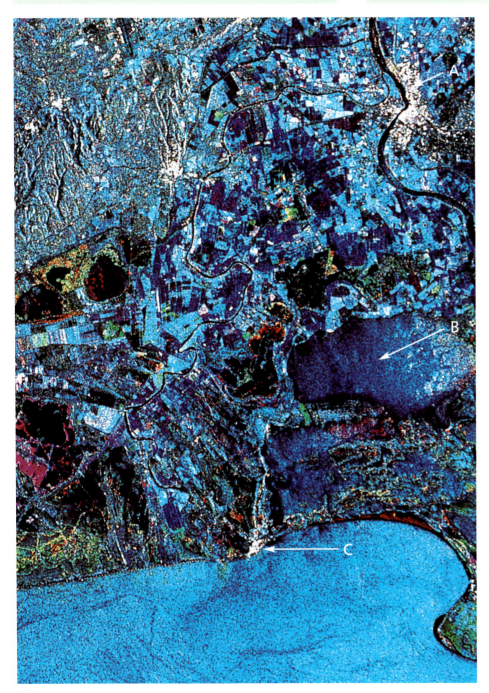

This is a false colour satellite image made by combining three photographs taken on different days over a period of four months. The white patches are towns. The blue colours show areas where crops and other vegetation have grown between the date of the first photograph and the date of the last photograph. The coast appears as dark green.
The Mediterranean Sea is at the bottom of the image.

2 Satellite image of the Rhône delta

Mapskills Units

2 Study Figure 3.
   (a) Write a few sentences describing the scene, to form a caption for the photograph.
   (b) Do you think this 'very special area' is worthy of careful management? Give reasons for your answer.

3 *White horses of the Camargue*

13 Rivers (3)

4 *Salt extraction in the Camargue*

3 Study Figures 5 and 6.
  (a) The approximate place where the photograph (Figure 5) was taken is shown on Figure 6. In what direction is the photograph looking?
  (b) Describe the shape of the fields in the lower half of the photograph.
  (c) Why do you think the fields are this shape?
  (d) Use the map to help you identify what forms the field boundaries. What is their purpose?
  (e) Use the map key to identify what is being grown in the fields.
  (f) Try to explain the different colours of the fields and suggest reasons why they are not all the same colour.
  (g) What is the river landform at X on Figure 5?
  (h) What is the name of the large lake at Y?
  (i) Heights above sea level are shown as spot heights and contours. The spot height values are in metres. What is the highest and lowest value on the map?
  (j) Write a few sentences to describe the landscape of the area.
  (k) Use the key to describe the vegetation on the map extract.

4 Study Figure 6.
Draw a sketch map of the western half of the map to show the course of the river as it approaches the sea (just visible at bottom left of the map). Include the Etang d'Icard, the fields of rice and the areas of woodland. Draw the D38 road. Complete your sketch by adding labels, a title, a scale and a north point.

Weblinks

Further details (in English) about the Camargue Regional Nature Park can be found at www.parcs-naturels-regionaux.tm.fr/lesparcs/camaa_en.html.

Mapskills Units

5 *Aerial view of the Camargue*

72

13 Rivers (3)

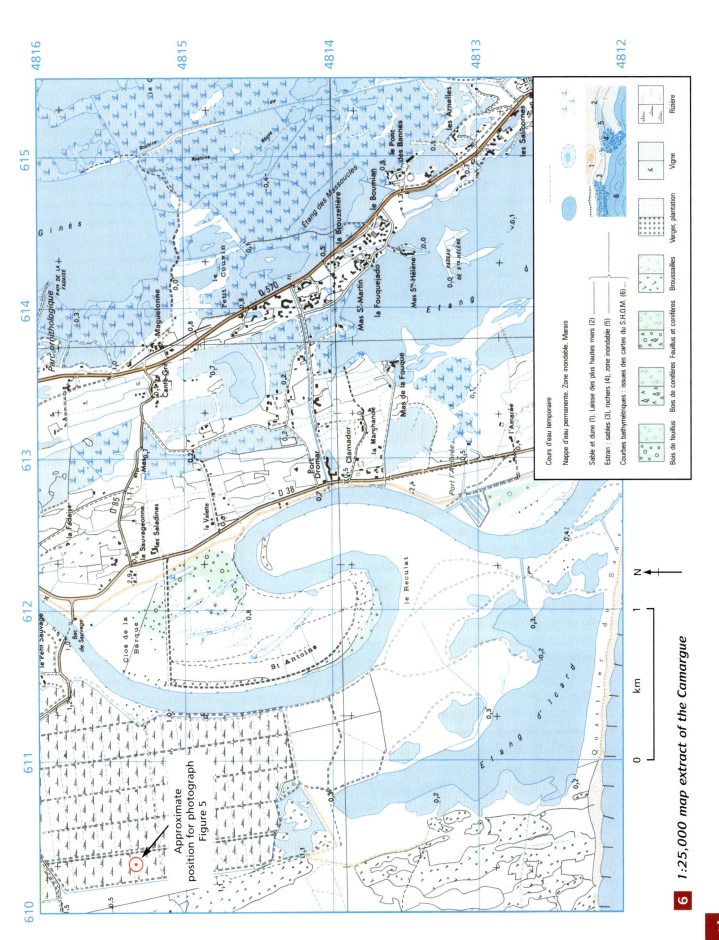

6 1:25,000 map extract of the Camargue

Mapskills Units

14 Climate

Comparing climates in Europe

The word *climate* means 'the average weather over a long period of time, usually 30 years'. It is different from *weather*, which describes the day-to-day conditions of the atmosphere, for example whether it is raining, if the sun is shining and what the temperature is.

Figure 1 shows that there is a wide variety of climates in Europe. In the far north conditions are very cold. This is because the sun is lower in the sky and less powerful than it is nearer to the Equator.

Notice also that there is a difference between western and eastern Europe. Towards the west, there is much more rainfall and temperatures are less extreme than they

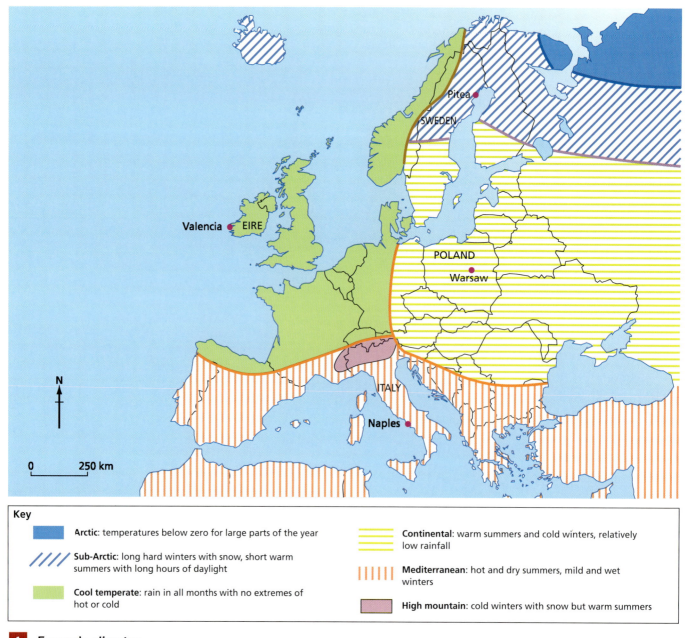

Key

- **Arctic**: temperatures below zero for large parts of the year
- **Sub-Arctic**: long hard winters with snow, short warm summers with long hours of daylight
- **Cool temperate**: rain in all months with no extremes of hot or cold
- **Continental**: warm summers and cold winters, relatively low rainfall
- **Mediterranean**: hot and dry summers, mild and wet winters
- **High mountain**: cold winters with snow but warm summers

1 *Europe's climates*

14 Climate

are in the heart of the continent. This is because the west is affected by the prevailing (most common) winds that come from the south-west. These winds bring lots of moisture (rain) and tend to be relatively mild throughout the year. Inland, well away from the sea, the moist Atlantic winds have less effect and the climate is drier and has greater extremes of temperature.

Climate has a huge effect on plants and animals and on human activities. Plants have adapted to cope with extremes of hot and cold and a lack of water. For example, orange trees have thick bark to help conserve water. Animals too have ways of coping. For example, in the Arctic, animals such as foxes and hares have thick fur coats to keep them warm, and they are white to help them blend into the snowy landscape. Human activities, particularly farming, reflect the climate. For example, in the Mediterranean climate, people grow fruit and vegetables that need hot, dry conditions, whereas in the wetter west, grassland and livestock-rearing dominates.

ACTIVITY

The aim of this activity is for you to make a comparison between two contrasting European climates. First you will need to choose one of the following contrasting pairs:

- North v South   Sub-Arctic (Sweden) and Mediterranean (Italy)
- West v East   Cool temperate (Eire) and Continental (Poland)

Study the photographs in Figure 3. Produce a poster or an ICT-based project report contrasting your chosen climate regions. You should aim to include the following on your poster:

- A map to show the two climates in Europe. Add names of countries, seas, etc.
- Climate graphs showing rainfall and temperatures, using the data in Figure 2 (see page 16 for help on drawing climate graphs).
- Brief written sections describing the extent of the two climate zones in Europe and contrasting the climates. You should also try to explain the differences.
- Some information about the way plants and animals have become adapted to living in the two climate zones (you will need to use the Internet and library facilities to help you).
- Some information about the effect of the climates on human activities. Look closely at the photographs and describe what they show.

Think carefully about how to design your poster so that it is effective when displayed on the wall. Make sure that you have a bold, clear title.

Sub-Arctic (Pitea, Sweden)

	Jan	Feb	Mar	Apr	May	June	July	Aug	Sept	Oct	Nov	Dec
Rainfall (mm)	37	25	23	28	30	47	50	68	69	48	48	44
Max. temp. (°C)	−6	−6	−1	5	11	17	21	19	13	6	0	−3
Min. temp.(°C)	−13	−14	−11	−4	2	8	12	10	5	0	−6	−10

Mediterranean (Naples, Italy)

	Jan	Feb	Mar	Apr	May	June	July	Aug	Sept	Oct	Nov	Dec
Rainfall (mm)	71	62	57	51	46	37	15	21	63	99	129	93
Max. temp.(°C)	11	13	15	19	23	28	30	30	26	22	16	13
Min. temp.(°C)	5	5	7	10	13	17	20	20	17	13	9	6

Cool temperate (Valencia, Eire)

	Jan	Feb	Mar	Apr	May	June	July	Aug	Sept	Oct	Nov	Dec
Rainfall (mm)	165	107	103	75	86	81	107	95	122	140	151	168
Max. temp.(°C)	9	9	11	13	15	17	18	18	17	14	12	10
Min. temp.(°C)	5	4	5	6	8	11	12	13	11	9	7	6

Continental (Warsaw, Poland)

	Jan	Feb	Mar	Apr	May	June	July	Aug	Sept	Oct	Nov	Dec
Rainfall (mm)	27	32	27	37	46	69	96	65	43	38	31	44
Max. temp.(°C)	0	0	6	12	20	23	24	23	19	13	6	2
Min. temp.(°C)	−6	−6	−2	3	9	12	15	14	10	5	1	−3

2 *Some climate figures*

Websites

Information about the weather and climate in Europe can be accessed at the Met Office's website at www.meto.govt.uk/index.html.

For information about animals and plants try a search. One good site for the sub-Arctic is the Canadian Wildlife Service at www.cws-scf.ec.gc.ca.

Mapskills Units

3 Climate conditions in different parts of Europe

76

15 Living in the mountains

Chamonix, French Alps

The Alps is one of the most spectacular mountain ranges in Europe. The highest mountain in Europe – Mont Blanc – is in the Alps. The Alps stretch across several southern European countries including France, Italy, Switzerland and Austria (see Figure 1).

Figure 2 shows part of the Alpine mountain range near Chamonix in France (Figure 5 on page 79 is a more detailed map of the area). Chamonix is a fairly small town but it is a very famous tourist resort and many people travel there each year to enjoy the mountains.

The Alpine landscape is spectacular and it offers opportunities for both summer sports, such as river rafting, mountain biking and trekking (Figure 3), and winter sports, such as skiing (Figure 4). Many towns and villages benefit from tourism. Visitors bring money to the shops, hotels and restaurants and jobs are created for local people.

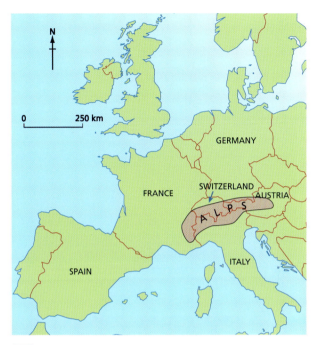

1 *Location of the Alps in Europe*

However, the cold winters, steep valleys and swollen rivers can cause enormous challenges for transport, farming and industry. Most settlements, farms and industry are squeezed onto the valley floors because the valley sides are so steep.

Engineers building roads and railways have had to cope with the hostile environment by cutting tunnels through the rock or building enormous viaducts. Winding roads with hairpin bends weave their way up mountainsides to form links between neighbouring valleys.

Farmers grow crops such as maize, vegetables and wheat on the fertile and warmer valley floors. On the higher slopes they graze cattle or sheep, bringing them down from the high pastures as winter approaches.

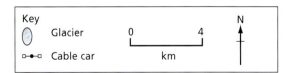

2 *1:200,000 map extract: Chamonix and Mont Blanc*

Mapskills Units

3

4   *Skiing in the Alps above Chamonix*

ACTIVITIES

1   Study Figure 2.
    (a) Locate Mont Blanc, the highest mountain in Europe. What is its height in metres above sea level?
    (b) What colour is used on the map to indicate high ground?
    (c) How are glaciers shown on the map? Give one example of a glacier.
    (d) Why do you think there are glaciers in the Alps?
    (e) Are there any main roads going over the Alps? Explain your answer.
    (f) How do people travel between the resorts of Chamonix and Courmayeur?
    (g) One way to travel across the Alps is by cable car via the Aiguille du Midi.
        • Draw the symbol used to show this route on the map.
        • How high is the Aiguille du Midi?

2   Study Figure 3.
    (a) At what time of year do you think this photograph was taken? Explain your answer.
    (b) Write a list of five or six words that describe the landscape. Now write a couple of sentences using your list of words to form a new caption to the photograph.
    (c) What are the people in the photograph doing?
    (d) Would you like to visit the area? Explain your answer.

3   Study Figure 4.
    (a) Why do you think that this area is one of the best resorts for skiing?
    (b) What type of vegetation is growing on the valley sides?
    (c) Suggest a possible hazard that might occur on the valley slopes as snow builds up.
    (d) How might snow cause problems for people living in Chamonix?

15 Living in the mountains

4 Study Figures 4 and 5.
   (a) Describe the site of Chamonix. (The 'site' is the point on the ground where a settlement has been built.)
   (b) Why do you think Chamonix has not expanded up the valley sides?
   (c) Apart from by road, how do you think tourists get to Chamonix?
   (d) What are the red lines that zig-zag their way up the valley sides?
   (e) Locate the N205 road in the south-west corner of the map extract. Follow its route in the direction of Chamonix. After about 1 km, just beyond les Pelerins, there is a fork in the road. The N205 now begins to head in a southerly direction towards the Mont Blanc Tunnel. Describe and suggest reasons for the course of this stretch of road. Draw a simple sketch map to illustrate your answer.
   (f) Look closely at the map and make a list of some of the things that tourists can do in Chamonix.

Weblinks

An excellent site offering information about Chamonix is at www.chamonix.net/english/home.htm.

Another good site is Compagnie des Guides de Chamonix at www.cieguides-chamonix.com/.

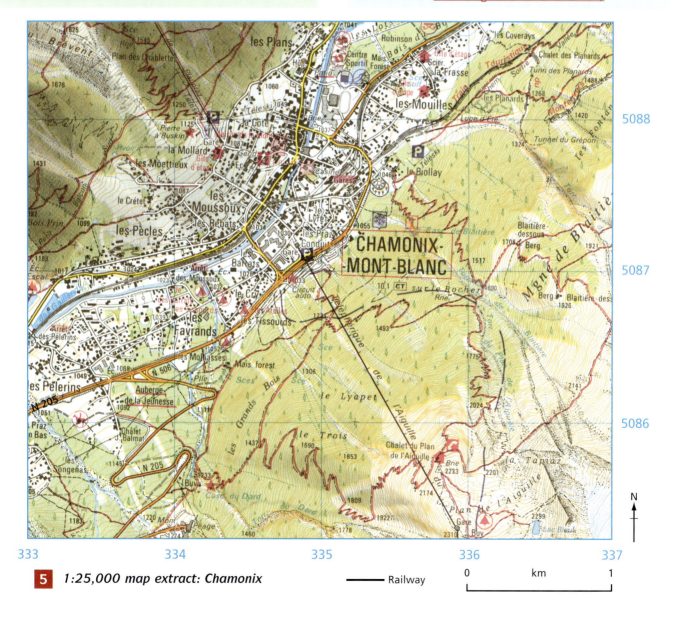

5 1:25,000 map extract: Chamonix ——— Railway

Mapskills Units

16 Industry

BMW car plant at Dingolfing, Germany

Figure 1 shows the huge BMW car plant on the outskirts of Dingolfing, Germany (Figure 2). This is BMW's largest car assembly plant. It was built on a 'greenfield' site in the early 1970s and currently employs about 23,000 people.

BMW carries out a number of different operations at Dingolfing. More than 1,300 cars are assembled here each day, mostly using computerised robots on the production line. In addition to the completed cars, the plant also produces kits for assembly in Asia, Africa and South America. Dingolfing is also BMW's main centre for the production of spare parts. Every day 90 trucks and 16 containers transport 30,000 parts to destinations all over the world.

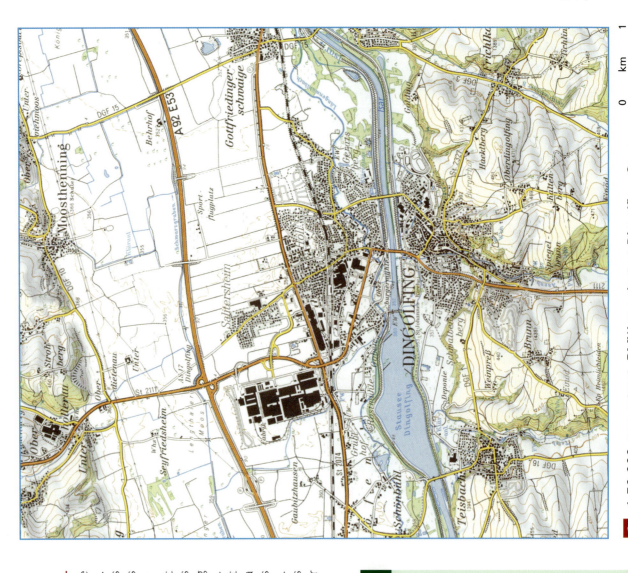

2 1:50,000 map extract: BMW car plant at Dingolfing, Germany

Activities

1 Study Figures 1 and 2. In Figure 2 the car plant is the collection of black, roughly square-shaped buildings to the north west of Dingolfing.

(a) In what direction is the photograph (Figure 1) looking?

(b) What is the name of the river in the photograph?

(c) What is the name of the wide river basin to the left of the car plant in the photograph?

(d) What is the number of the motorway?

16 Industry

1 BMW car plant at Dingolfing, Germany

More than 1,300 fully assembled cars are produced every day at this plant. In 2000 the 5 millionth BMW vehicle was produced here.

Mapskills Units

2 Study Figures 1 and 2.
   (a) What is the land surrounding the car plant mostly used for?
   (b) In the photograph Figure 3 was taken from approximately X on Figure 2. Why do you think trees have been planted in places around the car plant? Try to think of two reasons.
   (c) Use the scale on the map to attempt to calculate the area of land occupied by the car plant.

3 There were a number of reasons (called 'locational factors') why BMW chose to build a car plant on this site at Dingolfing. Use the map and photograph to explain how each of the following factors affected BMW's choice to locate here:
   (a) The car plant needed a huge flat area for buildings and car parks.
   (b) A large local labour force was needed.
   (c) Good road and rail links were needed to transport raw materials and the finished products.
   (d) Land was needed nearby for possible future expansion.

4 Study Figure 2. Draw a simple sketch map to show the location of the car plant. Show the main roads, the river, the railway and the town of Dingolfing. Add labels to indicate some of the locational factors listed in Activity 2. Include a scale, a north point and a title.

Websites

BMW maintains an excellent website (available in English) containing a great deal of information about the Dingolfing plant, at www.bmw-werk-dingolfing.de/.

Additional information about the BMW plant at Dingolfing is available at www.autointell.com/european_companies/BMW/bmw5.htm.

17 Quarrying

Glensanda 'superquarry', Scotland

Glensanda Quarry is on the west coast of Scotland in a very wild and remote part of the UK (Figure 1). It is one of only a few huge quarries (known as 'superquarries'), which the government hope will provide much of our future needs.

The rock that is quarried at Glensanda is granite. Granite is a very tough igneous rock that was formed many millions of years ago when molten magma cooled deep below the surface.

As Figure 1 shows, the quarry is located high up on a hill top. There are several rock faces (like cliffs) and explosives are used to blast away sections of rock. The broken rock, called aggregate, is transported from the quarry using a method called 'glory hole' mining. This involves transporting the rock on a conveyor belt through a hole in the quarry floor and then along a horizontal tunnel to a crushing plant alongside Loch Linnhe (Figure 3). The rock is then transported away from Glensanda by ship.

Gradually, the old quarry faces are landscaped so that they start to blend back into the hillside.

The quarry has been operational since 1986 and it produces about 6 million tonnes a year. Around 50% of the output goes to mainland Europe and 50% into England and Scotland via the ports of Glasgow, Liverpool and Southampton. This quarry is a long-term alternative to quarries in South West England and the Midlands providing road materials to South East England.

1 *Glensanda 'superquarry'*

Mapskills Units

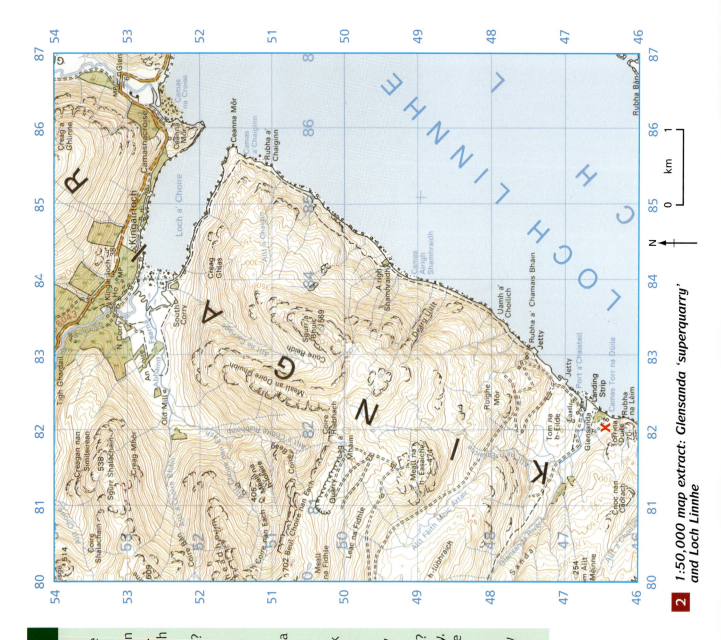

2 *1:50,000 map extract: Glensanda 'superquarry' and Loch Linnhe*

ACTIVITIES

1 Study Figures 2 and 3. The quarry is in grid square 8150. The lake (loch) in the photograph is Loch Linnhe.

(a) The photograph Figure 3 was taken from approximately X on Figure 2. In what direction is the photograph looking?

(b) What is the name of the inlet at A?

(c) According to the map, what type of tree is grown on the strips of land on the northern side of this inlet?

(d) Approximately how high above sea level is the quarry?

(e) Look at the photograph. What seems to be happening to the rock that arrives at the lochside from the quarry?

(f) What is the purpose of the narrow strip of land at B?

(g) What is the name of the river at C?

(h) Apart from the quarry and forestry, are there any other signs of people using the land?

(i) How does your answer to activity (h) help to explain why the quarry is here?

17 Quarrying

2 Study Figure 2. There is a small access road leading from Glensanda (8246) to the quarry. Find this on the map and follow its course. If the 'glory hole' method could not be used, the rock would have to be transported along this road. Suggest three problems of using this form of transport.

3 Study Figures 1 and 3.
(a) Why is the 'glory hole' method considered to be 'an environmentally friendly' technique for transporting the rock?
(b) Some people think 'superquarries' in remote areas such as Glensanda are very damaging to the environment, whereas other people think they are a good idea. Suggest some arguments in favour of and against 'superquarries'.
(c) Do you think 'superquarries' in remote areas are a good idea? Explain your answer.

3 Glensanda and Loch Linnhe

Weblinks

Glensanda Quarry is operated by Foster Yeoman. They have a useful website at www.foster-yeoman.co.uk.

A further source of information can be found at www.sams.ac.uk/dml/projects/reef/location.htm.

Mapskills Units

18 Volcano

Mount Etna, Italy

Mount Etna on the Italian island of Sicily is Europe's most active volcano. Most years an eruption takes place, sending huge clouds of ash high up into the atmosphere. As well as the ash, lava flows are common and whilst most do no serious damage, some do occasionally destroy farmland and buildings.

In October 2002 Etna erupted sending a thick plume of smoke and ash into the air (Figure 1). In a matter of hours, the ash was carried as far away as Libya in northern Africa, 560 km south of Mount Etna. Lava poured down the slopes of the mountain, wiping out orange groves and threatening to destroy houses. Some local schools were forced to close due to the threat of heavy falls of ash.

In July 2001, another major eruption threatened the tourist resort of Sapienza Hut (Figure 2) on the southern slopes of the mountain (Figure 3). Bulldozers worked to divert the lava away from the buildings.

1 *Mount Etna eruption, October 2002*

18 Volcano

ACTIVITIES

1. Study Figure 1. Answer the following questions with the help of an atlas.
   (a) In which direction is the wind blowing over Sicily?
   (b) What evidence is there of a change in wind direction over the southern Mediterranean?
   (c) Is the island of Malta being affected by the smoke and ash?
   (d) Is Tripoli, the capital of Libya, being affected by the smoke and ash?
   (e) From what direction would the wind be blowing if an eruption were to threaten the Greek capital Athens?

2. Study Figure 2.
   (a) Look at the lava flow at the top of the photograph. Write a few sentences describing what it looks like. (Is its surface smooth or bumpy? How do you know that it is hot? How thick is it compared with the height of the buildings?)
   (b) What evidence is there that the lava flow is not moving very fast?
   (c) Why do you think there are so many bulldozers, excavators and dumper trucks in the area?
   (d) Do you think that the machines could stop the flow of lava? Explain your answer.
   (e) Apart from the yellow machines, are there any other emergency vehicles present?
   (f) What are the arguments in favour of and against allowing small tourist resorts like Sapienza Hut to be built on the slopes of an active volcano?

3. Study Figure 3. Notice that all heights are given in metres above sea level.
   (a) What is the height of the main crater (Cratere Centrale) on Mount Etna?
   (b) Etna has more than one crater. How many are there?
   (c) Notice that the lava flows have been dated. Look around the map and find the flows for 1983. In which general direction did the lava flow in 1983?
   (d) Locate Sapienza Hut on Figure 3. Use the key to describe the uses of some of the buildings shown in the photograph Figure 2.
   (e) Where would you go close to Sapienza Hut to get a good view of the volcano? (Hint: you will need to look for the right symbol in the key.)
   (f) Make a list of some of the other attractions to tourists within about 3 km of Sapienza Hut.
   (g) Would you like to visit Mount Etna? Explain your answer.

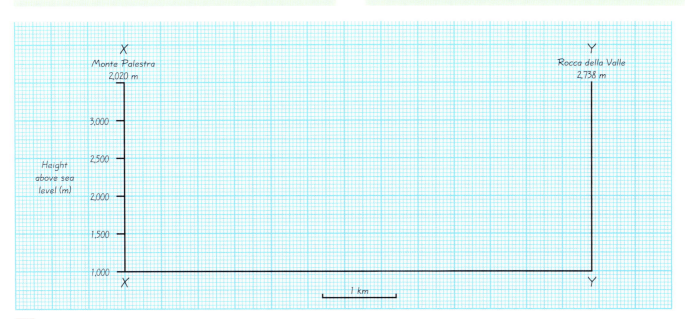

4 Axes for Activity 4

Mapskills Units

2 *Lava threatens Sapienza Hut, Mount Etna, July 2001*

4 Study Figure 3.
Draw a cross-section to show the shape of the volcano. (See pages 10-11 to remind you how to draw a cross-section.) You can choose your own line of section or use X–Y as indicated on Figure 3. A suggested scale for X–Y is shown in Figure 4. Don't forget to add labels (e.g. Mount Etna), axes labels and a title to your diagram.

Symbols used on Figure 3

- Hotel
- Restaurants
- Camping
- Starting point for excursions (planned)
- Guarded mountain hut
- Castle, fortress
- Ruins
- Panorama
- Cave
- Natural sights

Weblinks

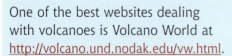

One of the best websites dealing with volcanoes is Volcano World at http://volcano.und.nodak.edu/vw.html.

There is a lot of information about Mount Etna, including some great photographs and satellite images, at http://volcano.und.nodak.edu/vwdocs/volc_images/img_etna.html.

A live Etna webcam can be accessed at www.vulcanoetna.it/it_etna_cam.php.

88

18 Volcano

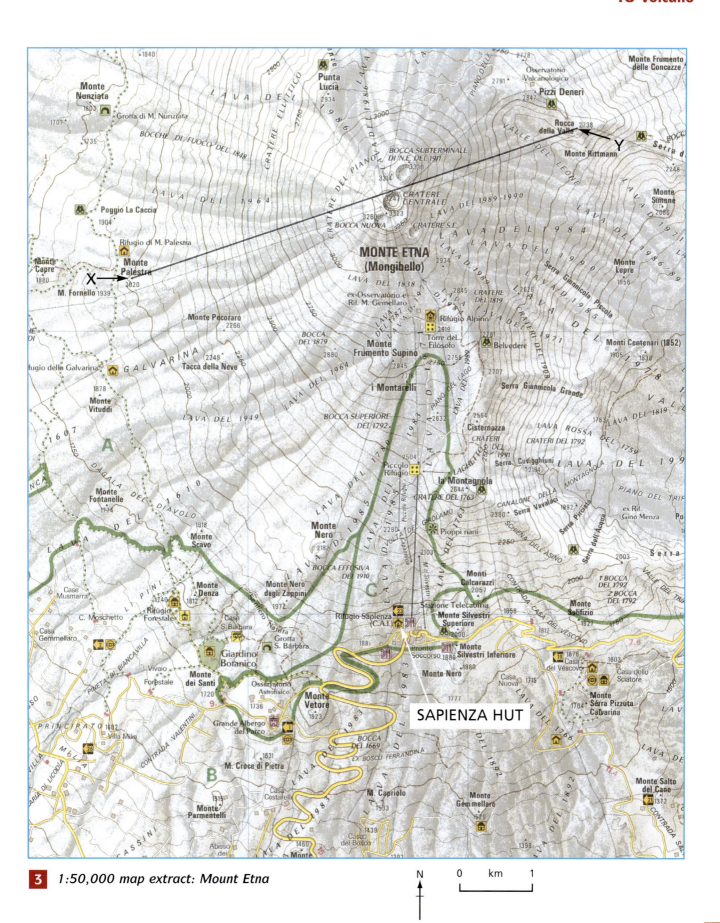

3 1:50,000 map extract: Mount Etna

Mapskills Units

19 Landslide

Karakoram Highway, Pakistan

A landslide is the collapse of part of a hillside. It can cause a great deal of damage and destruction to buildings and can result in people being killed and injured.

Landslides occur when part of a slope becomes unstable. It may become very wet after a period of heavy rain or it may be undercut by rivers or the sea. People can make slopes unstable by building on them or by cutting down trees whose roots normally help hold together soil and rock. An unstable slope collapses when a sudden event or 'trigger' occurs, such as an earthquake or a torrential rainstorm.

The Karakoram Highway connects Kashgar in China to Islamabad in Pakistan, over 1,250 km of the most rugged mountains in the world. It took 20 years to build and was opened in 1986. The map extract in Figure 2 shows the area very close to the Pakistan/China border.

Building and maintaining the Karakoram Highway has been a major challenge for engineers. Some 1,300 people died during the building of it. Much of the land is extremely steep (Figure 1) and the climate is harsh all year round. Snowmelt in the spring turns the rivers into raging torrents, and landslides, often triggered by small earthquakes, are extremely common. Several times a year, sections of the road either become blocked by landslides (Figure 3) or are simply washed away. Bulldozers patrol the road just to keep it open.

ACTIVITIES

1 Study Figure 1. It shows the typical landscape crossed by the Karakoram Highway. The road itself is in the bottom of the main Hunza valley (see Figure 2), although it does not show up clearly in the photograph.
   (a) Describe the steepness of the river valley sides.
   (b) Why do you think the valley sides are not farmed?
   (c) Look at the wide valley in the bottom left corner of the photograph. Is there any evidence that the valley floor is being farmed?

   (d) A small landslide has occurred at X. With the aid of a simple sketch, describe the shape of the landslide material that has collected at the foot of the slope.
   (e) Suggest how the following factors might have led to this landslide occurring:
      • steepness of the slope
      • undercutting by the river at the base of the slope.

2 Study Figure 2. This map is used by aircraft pilots navigating the very hazardous mountain range. It was produced in the USA and the height values are in feet rather than metres (3 feet = approximately 1 metre). The Karakoram Highway is the red line labelled 'K.H.' on the map.
   (a) What is the name of the river that runs alongside the Karakoram Highway?
   (b) Why do you think the road has been built in the valley bottom rather than on the valley side slopes?
   (c) Apart from landslides, suggest another natural hazard that might affect the road.
   (d) What do you notice about the locations of the sources of most of the rivers on the map?
   (e) What is the highest point on the map extract?

3 Figure 3 shows a small landslide that has come crashing down onto the Karakoram Highway. Vehicles are forced to either stop or take a hazardous detour that involves crossing a stream.

   You are a reporter for the Press Association. You have been asked to write a few sentences about the landslide shown in the photograph. Describe the nature of the landslide (how large are the boulders?) and the effect it is having on the traffic on the road. Try to suggest why the landslide may have happened (are the slopes steep?).

Weblinks

Additional information and some great photos can be found at
http://tours.hypermart.net/pakistan/kkh.htm and
www.pinlip.freeserve.co.uk/kkh.html.

19 Landslide

1 *Aerial view of Hunza River valley*

91

Mapskills Units

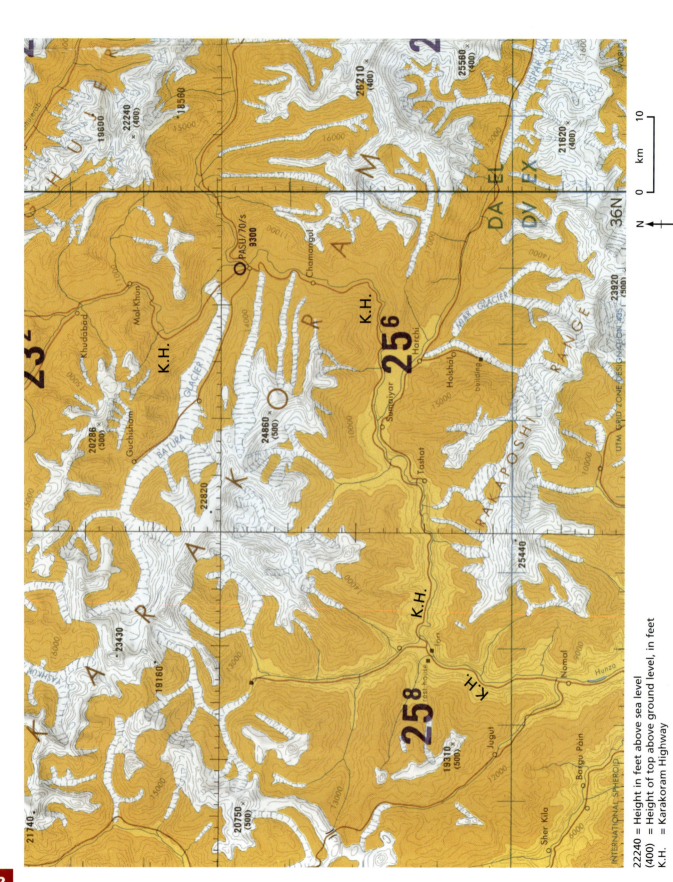

22240 = Height in feet above sea level
(400) = Height of top above ground level, in feet
K.H. = Karakoram Highway

2 *1:500,000 map extract: Karakoram Mountains and the Karakoram Highway*

19 Landslide

3 *A bus on the Karakoram Highway*

Mapskills Units

20 Population

The Earth at night

Figure 1 is a satellite photograph of the Earth at night. Notice that the Earth is far from being completely dark. Large regions appear white, which is the result of artificial lights from street lighting, industries and houses. The greatest concentration of artificial light is found in urban areas, such as in parts of Europe, North America and Japan. Those areas that appear black have no concentration of artificial light.

Figure 1 gives us quite an accurate impression of the distribution of the world's population. The highest population densities are found in Europe, eastern America and Japan and the lowest densities in North Africa, the Russian Federation, Canada and central parts of South America. Some areas, however, with high population densities, do not show up. This is particularly the case with countries in South-east Asia, such as India, China, Indonesia and the Philippines.

In addition to the areas of artificial light that appear white, the photograph also captures a couple of other interesting features. Over Greenland, for example, the very bright white band is the aurora borealis ('Northern Lights'), which occurs during the winter months in the Arctic. The aurora borealis is caused by radiation from the sun being affected by the Earth's magnetic field.

Notice also that there is a light orange band of colour over central Africa. This is the result of agricultural fires burning, clearing land before crops are planted.

ACTIVITIES

You may need to use an atlas to help you with the following activities.

1 Study Figure 1.
  (a) Is the greatest concentration of artificial light in the USA, in the east or in the west?
  (b) Name five European countries that have a high concentration of artificial light.
  (c) Name two countries in South America that produce large amounts of artificial light.
  (d) Locate Australia. The distribution of artificial light is a very accurate reflection of population distribution here. Describe the distribution of artificial light in Australia.
  (e) Name three large countries that produce little or no artificial light.

2 Study Figure 1.
  Try to name the cities labelled 1–10 on Figure 1.

3 Study Figure 1.
  (a) Use your atlas to suggest reasons for the lack of artificial light (and the low population densities) in areas A–D on Figure 1.
  (b) Use your atlas to suggest reasons for the large amounts of artificial light (and the high population densities) in areas E–G on Figure 1.

4 Study Figure 1.
  The satellite photograph does give quite an accurate view of the world's population distribution. However, some parts of the world such as India and Indonesia are known to have high population densities yet they do not show up as clearly as Europe and North America. Try to suggest reasons to account for this.

94

20 Population

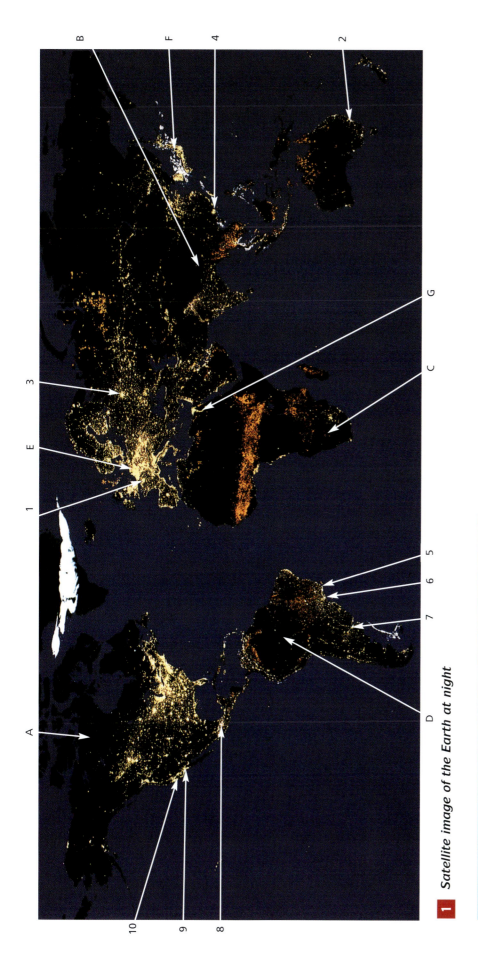

1 *Satellite image of the Earth at night*

Weblinks

One of the best sites for satellite photographs of the Earth is NASA's 'Visible Earth' website at http://visibleearth.nasa.gov/. Several interesting images can be downloaded to compare with Figure 1.

There are several sites offering satellite photographs of the Earth at night, including 'Angelfire' at www.angelfire.com/journal2/cp_lai/geosph.htm and the About Geography website at http://geography.miningco.com/cs/photos/.

Mapskills Units

21 Tourism

The Great Barrier Reef, Australia

The Great Barrier Reef is a maze of coral reefs stretching over 2,000 km along the Queensland coast of eastern Australia. Corals are living creatures that live inside tiny shell-like tubes each about 1 cm across. These tubes, rather like minute columns, are fixed together to form huge colonies of coral. It is these colonies that form the many thousand coral reefs that together make up the Great Barrier Reef.

The Great Barrier Reef is an amazing underwater world of magical caverns and more than 400 species of coral of all shapes, sizes and colours (Figure 1). It is a really special environment that attracts a great variety of fish. It is also home to many types of marine plants.

Coral needs warm and clear seawater to survive, which explains why reefs are found in tropical parts of the world. Coral is very delicate and is easily killed by pollution, such as oil spills, or by people walking on it.

Many people choose to visit coral reefs to go swimming, snorkelling or diving so that they can see the incredible wildlife. One of the most popular ports for people wishing to travel to the Great Barrier Reef is Cairns (Figure 2). Many companies run boat trips from Cairns Harbour to the reefs and islands, with Green Island (Figure 3) being one of the most popular destinations.

Activities

1 Study Figure 2.
 (a) What is the number of the main highway that runs through Cairns?
 (b) Apart from by road, how do you think most tourists travel to Cairns? Give evidence from the map to support your answer.
 (c) What is the straight-line distance between the sea front at Cairns and Green Island?
 (d) What is the compass direction from Cairns to Green Island?
 (e) Why do you think Green Island is one of the most popular reefs visited by people from Cairns?
 (f) Can tourists stay overnight on Green Island?
 (g) Another very popular destination from Cairns is Fitzroy Island. Describe the route taken by boats from Cairns to Fitzroy Island.
 (h) What are the attractions of Fitzroy Island to tourists?
 (i) Why are people warned not to go swimming in 'Far Northern rivers and streams'?
 (j) Why is swimming dangerous in the sea during the summer?

1 Great Barrier Reef coral garden

21 Tourism

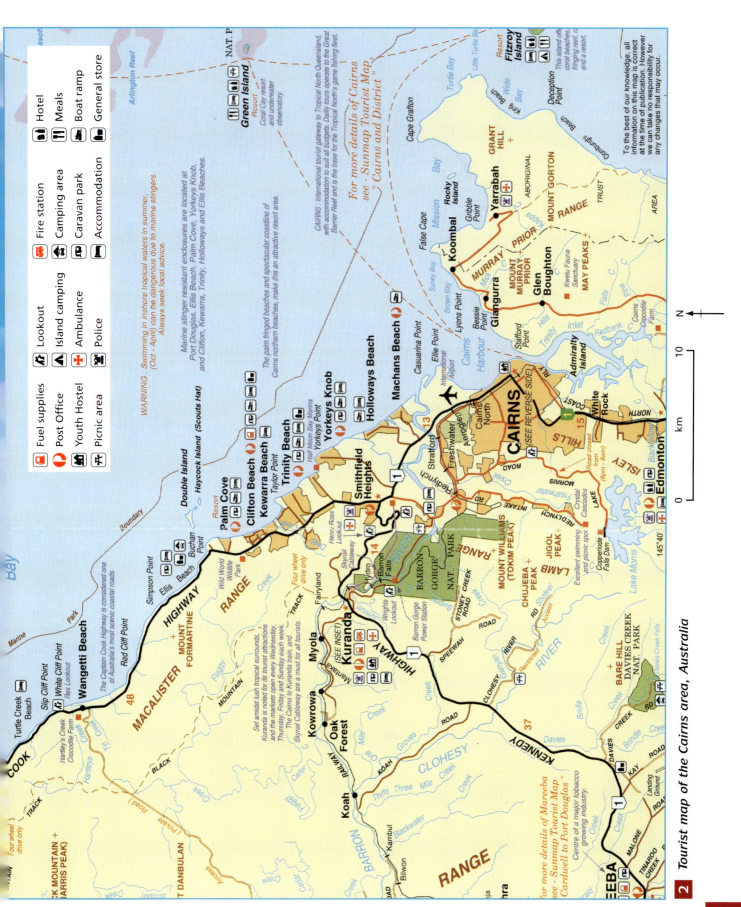

2 Tourist map of the Cairns area, Australia

Mapskills Units

3 *Green Island*

2 Study Figure 2.
   You and your family are spending a few days in Cairns on holiday. It is a windy day and boat trips have been cancelled because the sea is too rough. Use the map to suggest what to do and see on the land. Describe where you would go and why. Draw a simple sketch map to show your route from Cairns and the main attractions that you intend to visit. Don't try to do too much!

3 Study Figures 2 and 3.
   (a) Use the photograph to describe what Green Island is like.
   (b) Is Green Island somewhere that you would like to visit? Explain your answer.
   (c) Do you think people should be allowed to build hotels and resorts on islands like Green Island? Explain your answer.

Weblinks

There are many sites offering information about the Great Barrier Reef.

The Great Barrier Reef Marine Park Authority site at www.gbrmpa.gov.au gives lots of background information about the reef.

National Geographic has a virtual dive at www.nationalgeographic.com/earthpulse/reef/reefl_flash.html.

Further information about Green Island and the Green Island Resort can be found at www.greenislandresort.com.au/.

22 Water supply

Las Vegas, Nevada, USA

Las Vegas is a large city in the state of Nevada, USA. It has a population of over 483,000 and is part of a huge urban area, which together houses some 1.4 million people. Situated in a desert, one of the major issues facing the city's authorities is water supply.

The desert climate of Las Vegas means that there is very little rainfall and cloud cover. This results in high summer temperatures but quite cold winters.

About 88% of Las Vegas' water comes from Lake Mead on the Colorado River (Figure 1). The remaining 12% of Las Vegas' water is pumped to the surface from groundwater supplies up to 500 m deep.

Las Vegas is an extremely popular tourist resort with a huge number of clubs and attractions. The main tourist area is called 'The Strip' (see Figures 2 and 3). The demand for water is enormous and the local authorities are very concerned about being able to supply enough to satisfy the city's needs in the future.

There are several resorts that make use of water as part of their attractions, for example 'Wet N Wild Water Park' on Boulevard Las Vegas (Figure 2). However, these only account for 7% of all the water used. In comparison, residents in Las Vegas use 65% of all the water. Outdoor water for landscaping alone (e.g. for watering lawns) accounts for 70% of total residential water use.

Whilst it is hoped to be able to take more water from the Colorado and also to recycle more 'used' water, a lot of pressure is being put on people to use water more wisely. This is called **water conservation**. A target of 25% conservation has been set for the year 2010.

1 Las Vegas, Nevada

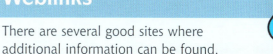

Weblinks

There are several good sites where additional information can be found.

The Las Vegas Convention and Visitors Authority (www.lasvegas24hours.com) has a great deal of information about Las Vegas, including population data.

For further information on water supply, access the Southern Nevada Water Authority at www.snwa.com.

Maps of Las Vegas and its surroundings can be obtained from Las Vegas Maps at www.lasvegasmaps.com.

Mapskills Units

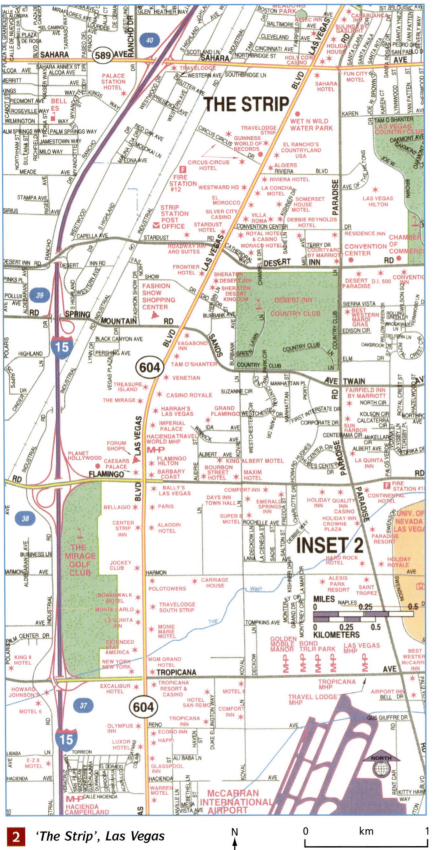

2 'The Strip', Las Vegas

Activities

1. Study Figure 2.
   (a) What number road is Boulevard Las Vegas?
   (b) Make a list of five tourist attractions on the map.
   (c) There are lots of hotels in central Las Vegas. What do you think would be the main demands for water in a hotel?
   (d) Why are large quantities of water needed to maintain the golf courses shown on the map?
   (e) What is the name of the Las Vegas international airport?
   (f) Suggest the uses for water at the airport.

2. Study Figures 2 and 3. The photograph in Figure 3 is looking south along the Boulevard Las Vegas. The Paris casino/hotel at the bottom of the photograph is on the opposite side of the Boulevard Las Vegas to the Mirage Golf Club (Figure 2).

Symbol	Meaning
1 (green)	Golf course
(tree)	Forest
	Cemetery
	College/University
	Shopping centre
	School
•	Point of interest
H	Hospital
F	Fire station
MHP	Mobile home park
✱	Casino/Hotel
■	Government building
P	Police station
▲	Library
✉	Post office

22 Water supply

3 Las Vegas

(a) What is the name of the road that joins the Boulevard Las Vegas at X on Figure 3?
(b) What evidence is there on the photograph that the area receives very little rainfall?
(c) Where on the photograph would you expect outside landscape watering to take place?
(d) It is often said that American cities are designed for cars. Do you think this saying is true for Las Vegas? Explain your answer.

3 Figure 4 gives population figures for Las Vegas from 1970 to 2000.
(a) Plot these figures on a line graph, with the date along the bottom (X) axis and the population along the side (Y) axis.
(b) Describe the trend of population shown by your graph.
(c) How does the trend present a challenge for water supply in the future?

1970	1980	1990	1996	1997	1998	1999	2000
125,787	164,674	268,330	398,110	425,270	441,230	465,050	483,448

4 Population of Las Vegas, 1970–2000

Mapskills Units

23 LEDC city

Mexico City, Mexico

Mexico is a large country in Central America. It has a population of 97 million people, which is twice the population of England. About 95% of the people speak Spanish.

Mexico City is the capital city and, with a population of 17 million, it is one of the largest and fastest-growing cities in the world. The site of the city is a vast shallow bowl encircled by volcanic peaks (Figure 2). It was once a lake-bed that has now dried up. The mountains restrict the outward growth of the city, which is why the buildings are packed closely together.

Life in the city is fast and furious. There are traffic jams on the main roads, which can last most of the day, and car horns, buses, police-whistles and people contribute to the high levels of noise in the city. The air is heavily polluted as a result mainly of the traffic, and it leaves many people with a metallic taste in the mouth and a sore throat.

The centre of Mexico City is colourful, vibrant and exciting. There are many street markets where local people sell freshly made take-away food, drinks, fruit and vegetables, and arts and crafts (Figure 1). Many people are involved in the 'informal sector', buying and selling to make money, because in a city with so many people there is a shortage of well-paid full-time jobs. Mexico City is home to several magnificent buildings, museums and art galleries and there are also attractive parks.

ACTIVITIES

1 Study Figure 1.
   (a) The yellow VW 'Beetle' cars are a common sight in Mexico City. What do you think these cars are used for?
   (b) Suggest two reasons why the market stalls are covered.
   (c) Look closely at the market stalls and make a list of the types of items being sold (e.g. food, DIY, crafts).
   (d) Do you think the market is mostly aimed at local people or tourists? Explain your answer.
   (e) Look at the buildings behind the market stalls. Are most of them one, two or three-storey buildings?
   (f) Do the buildings look relatively old or modern? Explain your answer.

2 Study Figures 2 and 3. The photograph in Figure 2 is taken looking north along the Avienda Lazaro Cardenas (this main road runs north–south through the centre of the city). The grand building at the bottom of the photograph is the Palacio de Bellas Artes.
   (a) Use the map to locate and name the following features shown on the photograph:
   • the name of the park at A
   • the name of the road at B
   • the function of the building at C
   • the name of the major urban freeway (motorway) at D.
   (b) Look at the key in Figure 3. What symbols are used to show the underground, and the underground stations?
   (c) You have just visited the Palacio de Bellas Artes and have walked to the Bellas Artes underground station. You want to visit the main square (Zocalo) in Mexico City. Do you take a train going east or west? How many stations do you go through before reaching Zocalo?
   (d) Make a list of some of the attractions and facilities for tourists in the centre of Mexico City. (You may need to refer to a Spanish dictionary!)

23 LEDC city

3   Study Figure 3.
   (a) Describe the layout of the roads on the map extract.
   (b) Do you think the centre of Mexico City has been well planned? Explain your answer.
   (c) Use the key to suggest what has been done to manage traffic in the centre of the city.
   (d) What other forms of management could be used in a city like Mexico City to reduce traffic and traffic pollution?

Weblinks

An excellent site, which includes a virtual tour of Mexico City and many good links, is at www.mexicocity.com.mx/mexcity.html. There are several sites offering information for tourists, which will be revealed by a standard search.

1 *A street in the centre of Mexico City*

Mapskills Units

2 *Mexico City*

23 LEDC city

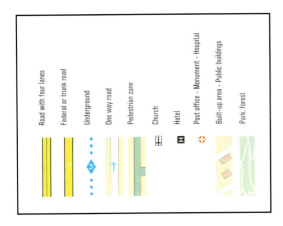

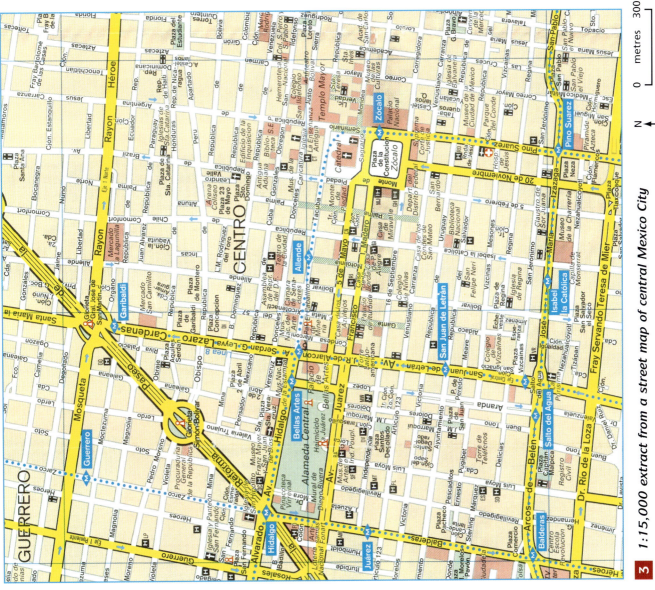

3 1:15,000 extract from a street map of central Mexico City

105

24 Tropical rainforest

Pasoh Forest Reserve, Malaysia

Tropical rainforests are found in a broad belt stretching from Central and South America, through Africa and into South-east Asia. They are truly wonderful natural environments, which boast more species of plants and animals than any other environment on Earth. They are also home to 200 million tribal people.

Tropical rainforests have been cut down to make way for farming, logging and extraction of raw materials, such as iron and copper. In some countries huge areas of natural (virgin) forest have been destroyed to make way for commercial enterprises. Across the world, conservationists are trying to slow down the rate of deforestation. This is because plant and animal species are being threatened with extinction and local people are in danger of losing their homes and livelihoods. There is also concern that the removal of forests may lead to the climate becoming drier, as less water will be returned to the atmosphere by the process of transpiration.

Malaysia (Figure 1) is a country right in the heart of the tropical rainforest belt and, in the past, most of the country was rainforest. Now a little over 50% of the virgin forest (Figure 2) remains, and the government is keen to reduce the amount of destruction, particularly by illegal loggers.

About 70 km to the south-east of the Malaysian capital, Kuala Lumpur, is Pasoh Forest Reserve (Figure 3). This area of forest is managed very carefully to enable some commercial logging to take place but also to conserve areas of natural rainforest. At the heart of the reserve is an area of virgin rainforest (Figure 2). This is surrounded by previously cut forest that is now being allowed to re-grow. Scientists based at a local research station monitor the rainforest. They are very interested to see how the area deforested in the 1950s (Figure 3) regenerates, as this could help in other parts of the world where rainforest destruction has occurred.

ACTIVITIES

1 Study Figure 2. This photograph was taken at a height of about 40 m above the forest floor looking out over the tops (canopy) of the forest.
   (a) Are all the trees the same height?
   (b) Do the trees seem to be very closely spaced or are there large gaps between them?
   (c) Do you think it would be light or dark if you were standing on the forest floor? Explain your answer.
   (d) In the tropics, the climate is wet and warm. How does this explain why there are so many trees?

1 *Malaysia*

24 Tropical rainforest

2 The canopy of the tropical rainforest

2 Study Figure 3.
  (a) What is the approximate distance from north to south of the Reserve?
  (b) What is the approximate distance from west to east?
  (c) Multiply your answers to (a) and (b) to discover the approximate area of the Reserve.
  (d) Describe the location of the regenerating forest in relation to the virgin forest.
  (e) What do you think would happen to the virgin forest if it were not protected?
  (f) Do you think it is important to protect areas of virgin rainforest? Explain your answer.
  (g) Why do you think there are no extensive forests on the land over 300 m?

3 Study Figures 3 and 4.
  (a) What is the height (in feet above sea level) of the highest peak in the Reserve?
  (b) Is the small town of Simpang Pertang connected to the main electricity supply line?
  (c) Apart from road, what other form of transport might be used to take away timber and crops?
  (d) Locate Kpg Tengah at A on Figure 4. Describe the pattern of the roads here. Suggest possible reasons for this layout.
  (e) What evidence is there on the map that areas of forest have been cut down? You will need to look very carefully!
  (f) Locate area B on Figure 4. This area of cleared land is now being used for growing rice. What advantage does this area have for the production of rice?

Mapskills Units

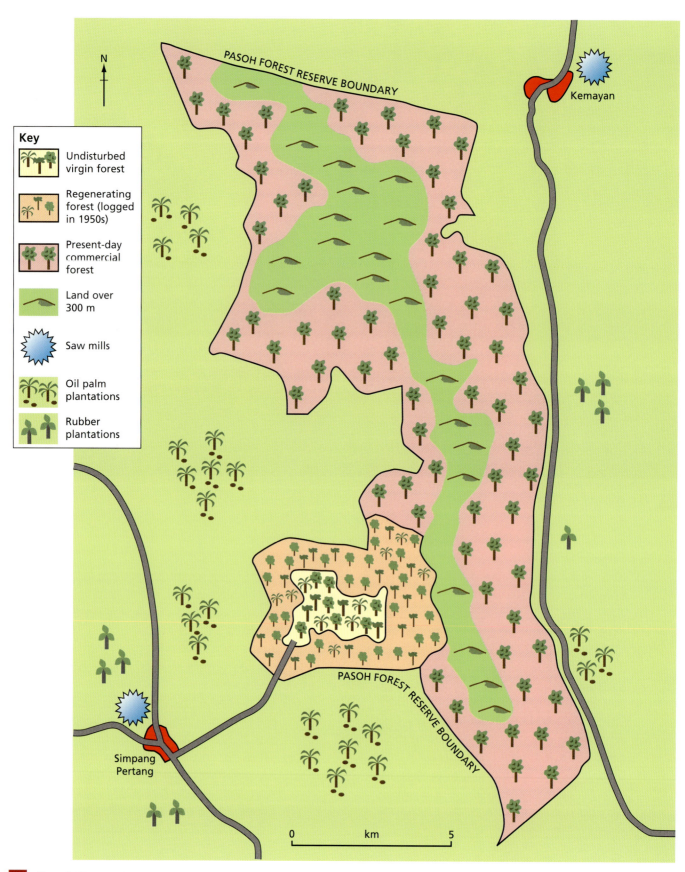

3 *Pasoh Forest Reserve, Malaysia*

24 Tropical rainforest

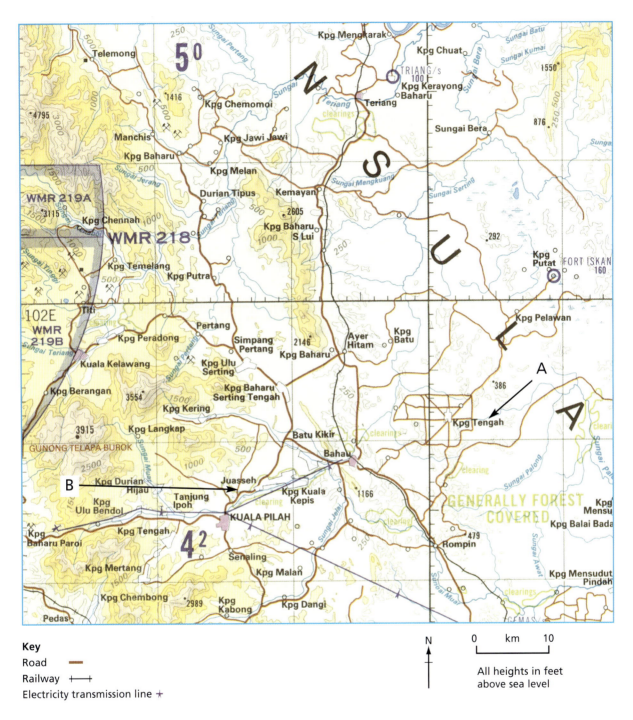

Key
Road ▬
Railway ┝━┥
Electricity transmission line ✱

4 1:500,000 map extract: *Pasoh Forest Reserve*

All heights in feet above sea level

Mapskills Units

5 *The saw mill in Pasoh Forest Reserve*

4 Study Figures 3 and 5.
  (a) What does a saw mill do?
  (b) How many saw mills are there in the area and where are they located?
  (c) How do you think the timber is transported to and from the saw mills?
  (d) Why are the saw mills located so close to the forest?

5 Study Figures 3, 6 and 7. A plantation is an intensive form of farming involving growing cash crops. In Malaysia, oil palm and rubber are common plantation trees.
  (a) Are the plantations located inside or outside the Reserve?
  (b) Describe the layout of the rubber trees in Figure 6.
  (c) Use Figure 7 to describe how rubber is collected from the trees.
  (d) Try to list five everyday products that are made from rubber.

24 Tropical rainforest

6 *A rubber plantation*

Weblinks

Much of the material, including the photographs, used in this unit come from an excellent publication *Tropical Rainforests* published by the Field Studies Council. The FSC website is at www.field-studies-council.org/.

The Malaysian Timber Council at www.mtc.com.my/ has some useful articles and statistics.

The Ministry of Primary Industries in Malaysia at www.kpu.gov.my/commodity/comodity.htm has good background information on agricultural products including rubber and oil palm.

An excellent site about the research at Pasoh can be found at www.frim.gov.my/pasoh1.html.

7 *Collecting rubber from a tree*

111

Key Stage 3 Assessment Tasks

A Locating a new hypermarket

Aylesbury, Buckinghamshire

This assessment will provide you with a National Curriculum level for Geography. Study all of the information on this page and on pages 114–117. The Activities on this page state what you need to do to complete the assessment.

You also need to read the mark scheme on page 113 before you start. This is what your teacher will use to give your work a National Curriculum level. Ask your teacher to explain what each different level means. It is important for you to know what you have to do in order to succeed, and how to achieve each level. Remember that in order to succeed, you must always **explain** and **give reasons for** any choices that you make.

ACTIVITIES

Imagine that you work for the local council in Aylesbury, a town in Buckinghamshire. Your task is to recommend the best location for a new hypermarket to be built somewhere in the town. You are provided with background information and techniques to use in order to help you reach your decision (Figures 1–4). Your assessment should take the form of a report to the Chairperson of the local council, and should include the following:
1. An explanation of why Aylesbury needs a new hypermarket.
2. A completed copy of Figure 4.
3. Your chosen location. You need to **justify** (give reasons for) your choice. Use all of the information provided in Figures 1–4 to help you. You may wish to consider factors other than those in Figure 4.
4. A brief explanation of the reasons why you did not choose each of the other sites.
5. A large advertisement suitable for the local evening paper, to explain your decision to the people of Aylesbury.

Aylesbury is the county town of Buckinghamshire. It was a small town until the end of the 19th century when the completion of the railway line to Manchester led to a rapid growth of industry and population.

During the 20th century, Aylesbury's main function was that of a market town serving the surrounding area. In the 1960s, the town centre was redeveloped to make it more attractive.

Since then, its location within easy reach of London has led to a further rise in both population and prosperity. Aylesbury is now a commercially successful town of over 60,000 people. The population has doubled in the last thirty years, and is still increasing. The prediction of continued growth has led the council to plan more building, including hypermarkets, new housing and an industrial estate.

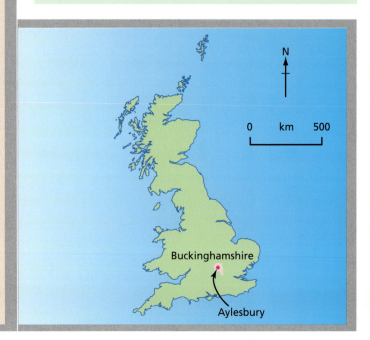

1 The location of Aylesbury

A Locating a new hypermarket

Mark Scheme

Level 3

- You describe the different places. You also compare them.
- You know that different places have the same and different features.
- You give some reasons for your descriptions.
- You use some skills and evidence to answer the questions. Some of the vocabulary you use is accurate.

Level 4

- You describe and compare different locations.
- You describe what causes these places to be different.
- You show how people can change places and affect the lives of people living there.
- You explain your own views, using a range of skills and evidence. Your use of vocabulary is quite accurate.

Level 5

- You describe and begin to explain processes caused by people.
- You give reasons for the ways in which people change environments, and the different views people have about them.
- You explain your own views and opinions, using appropriate skills and methods of presenting information.
- You make sensible conclusions as a result of your work.

Level 6

- You describe and explain relevant processes caused by people. You recognise that these processes interact to produce the characteristic features of a location.
- You understand how conflicting demands upon a place may occur.
- You appreciate that different values and attitudes have an impact upon places and environments.
- You clearly explain your own views and judgements.
- You present your findings clearly, and reach appropriate conclusions.

Level 7

- You describe how physical and human processes are linked, and how this affects geographical patterns and helps change places.
- You understand that many factors, including values and attitudes, influence the decisions made about places.
- You clearly explain your own judgements and choices. You relate this to information other than that provided for this assessment.
- You present well-argued summaries and reach detailed conclusions.

Level 8

- You offer explanations for interactions between physical and human processes that give rise to change in places and environments.
- You use your knowledge and understanding to explain the linkage of these factors in causing change in the location studied.
- You begin to evaluate the relative importance of factors, including people's values and attitudes, in affecting change and development.
- You are able to justify your own judgements, and begin to evaluate sources of evidence used.
- You present full and coherently argued summaries of your investigations and reach detailed conclusions.

Exceptional performance

- You explain complex interactions within and between physical and human processes that give rise to the characteristics of places and the nature of change.
- You refer specifically to geographical factors to explain and predict change in the characteristics of places over time.
- You draw selectively on geographical ideas and theories appropriate to the investigation to back up your opinion, judgements and conclusions.
- You use a wide range of skills and the full range of available evidence in your investigation.
- You evaluate critically sources of evidence and present coherent arguments and effective, accurate and detailed conclusions.

Key Stage A

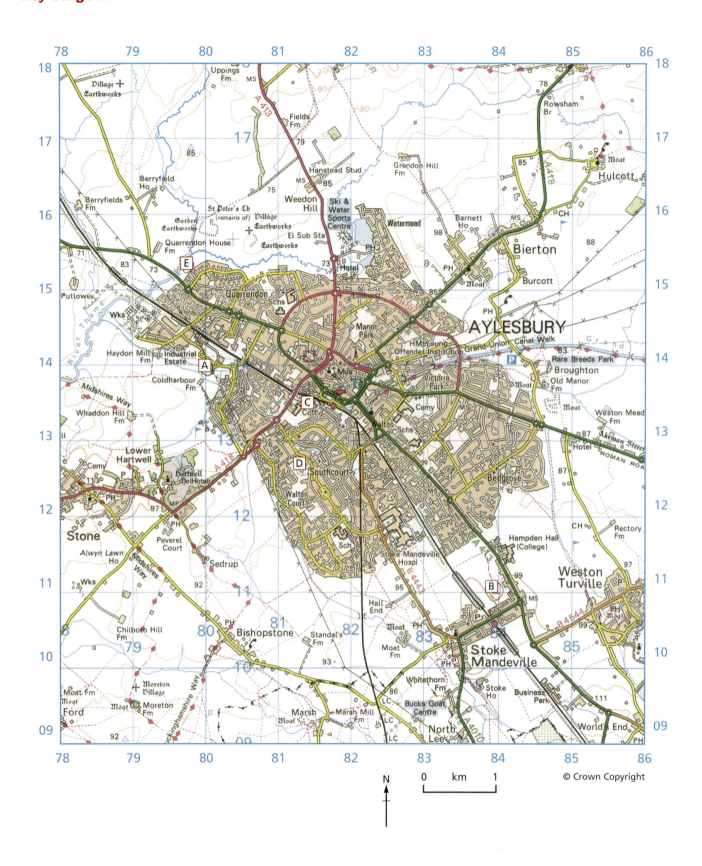

2 *1:50,000 OS map extract: Aylesbury in Buckinghamshire*

A Locating a new hypermarket

Where could we locate the new hypermarket in Aylesbury?

A Rabans Lane
On the edge of a new development area, which includes both housing and industry. This is likely to expand further, but it is near to a tributary of the River Thame and private farmland.

B Stoke Mandeville
Open land, near to the railway and main road, but separate from the main urban area of Aylesbury.

C Aylesbury College
Near the town centre and backing onto the railway line. The College may wish to sell some of its land in the future, for commercial development.

3 *Five possible locations for a hypermarket in Aylesbury*

Key Stage 3 Assessment Tasks

D Walton Court

In the centre of a 1960s suburb. Some derelict and open land.

E Quarrendon

On a main road into Aylesbury and near to established housing. On the floodplain of the River Thame.

A Locating a new hypermarket

Factor	Site A	Site B	Site C	Site D	Site E
Accessibility					
Cost of land					
Suitability of land					
Market					
Objections					
TOTAL SCORE					

THE FACTORS

Accessibility: Will the hypermarket be easy to get to, for both customers and deliveries?

Cost of land: Is the land likely to be very expensive, or quite cheap? A hypermarket will need lots of land for the store and for the car park.

Suitability of land: Can the land be used easily? Does it, for example, need to be cleared of other land uses? Is it likely to flood and therefore need flood defences?

Market: Will there be a large number of people living locally to use the hypermarket and make it profitable?

Objections: Could there be some people who would be against the building of a new hypermarket here?

4 *Matrix: Choosing the best location for a hypermarket*

For each factor, give each location a score from 1 to 5. A score of 5 means that the location is excellent, 1 means it is unsuitable.
Calculate the total score for each location.
The best location is the one with the highest total.

Key Stage 3 Assessment Tasks

B Living with earthquakes

California, USA

This assessment will provide you with a National Curriculum level for Geography. Study all of the information on this page and on pages 120–122. The Activities on this page state what you need to do to complete the assessment.

You also need to read the mark scheme on page 119 before you start. This is what your teacher will use to give your work a National Curriculum level. Ask your teacher to explain what each different level means. It is important

for you to know what you have to do in order to succeed, and how to achieve each level. Remember that in order to succeed, you must always **explain** and **give reasons for** any choices that you make.

This assessment is about earthquakes. As well as Figures 1–7 on pages 118–122, you will need to use your knowledge and understanding of earthquakes from your Key Stage 3 Geography lessons.

ACTIVITIES

- Imagine that you work for a British newspaper. You have been asked to produce an article for the colour section of the newspaper's Sunday edition.
- The title of your report is 'Living with Earthquakes'.
- Your completed news report should include the following sections.
  1 A brief description of where earthquakes occur. You could use a map to illustrate your answer.
  2 An explanation of the distribution of earthquakes.
  3 A report on the ways in which people cope with earthquakes. This should include how to prepare for an earthquake, and how to cope with its effects.
  4 An explanation of why a major earthquake is expected to hit San Francisco within the next thirty years.

5 A section headed 'Preparing for the big one in San Francisco', to include the following:
  (a) The chances of an earthquake occurring.
  (b) A description of the areas and building structures at greatest risk from the effects of an earthquake.
  (c) Your views on the best location for two new emergency facilities. The authorities have decided that, with an increasing population, they need to build a new hospital and a new fire station. Explain where you think they should be built.

- You may produce your report in any format. It may be hand-written or word-processed. Try to use a variety of methods of presenting your information. Remember that you are being assessed on your Geography. The content of your work is more important than the presentation.

What is an earthquake?

An earthquake is the sudden shaking of the ground that follows a release of energy in the Earth's crust. This is usually caused by movements at the boundaries of the Earth's tectonic plates (see Figure 2). Earthquakes may also be caused by volcanic eruptions.

1 *About earthquakes*

B Living with earthquakes

Mark scheme

Level 3

- You describe where earthquakes are found, but don't explain this very well.
- Some of the advice you give about earthquakes may be right, but some is wrong or not properly finished.
- You do not describe the chances of future earthquakes very accurately.
- You make suggestions for where to locate the new hospital and fire station, but these may not be very suitable.

Level 4

- You describe where earthquakes are found, with some explanation.
- Some of your earthquake advice is good, but it varies in detail and accuracy.
- There is some accuracy in your prediction of future risk and the location of the hospital and fire station.

Level 5

- You describe the location of earthquakes in detail, with good explanation.
- Your earthquake advice is clearly communicated, and mostly relevant.
- The descriptions of future risk are accurate, and the location of the emergency facilities is appropriate.

Level 6

- You describe and explain the location of earthquakes, making some reference to plate tectonics.
- Your earthquake advice is relevant and accurate. You consider the difference between short-term and long-term needs.
- The description of future risk is accurate, and uses references from the map.
- The location of emergency facilities is appropriate. You give reasons for your choices.

Level 7

- You make a thorough and detailed description and explanation of the location of earthquakes. This is clearly linked to plate tectonic theory.
- Your earthquake advice is appropriate. It considers the needs of people, and refers to both the short-term and long-term effects of earthquakes.
- The description of future risk is accurate, it refers to locations on the map, and makes comparisons between different locations.
- The location of emergency facilities is appropriate and clearly justified.

Level 8

- You clearly describe the distribution of earthquakes. You make good use of geographical terms, and include reference to plate tectonics.
- Your earthquake advice is excellent, and you think about the needs of different people. You understand the difference between short-term and long-term effects.
- You accurately describe the chances of a future earthquake. You compare places located on the map in order to do this.
- Your suggested locations for new emergency facilities are appropriate, and you give detailed reasons for the locations you have chosen.

Exceptional performance

- Your description of the pattern of earthquakes is coherent and integrated with the explanation. It includes located examples, with reference to plate tectonic theory to support this section.
- Earthquake advice is entirely relevant, considers the needs of different groups of people, and includes relevant place-specific advice.
- There is a detailed description of future risk, using located examples and making reference to the growth of population in the area.
- The location of future emergency facilities is appropriate and fully justified. There is reference to the integration of such facilities into the wider network of protection measures.

Key Stage 3 Assessment Tasks

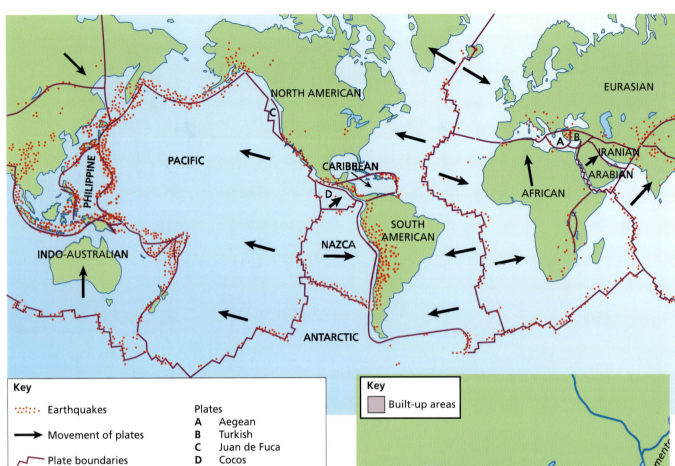

2 *World plate boundaries and major earthquakes*

MEMO:

Major quake likely to strike San Francisco between 2000 and 2030

On the basis of research since the 1989 San Francisco earthquake, scientists conclude that there is a 70% chance of a major earthquake hitting the region before 2030. Such an earthquake, which would measure 6.7 or more on the Richter scale, would cause extensive damage (see Figure 5). Authorities are concerned that an earthquake could occur at any time in this region of rapidly growing population. This shows the urgency for all people living in the area to continue preparing for earthquakes.

3 *Extract from the United States Geological Survey*

4 *Location of San Francisco*

B Living with earthquakes

5 *The damage caused by the 1989 earthquake in San Francisco*

121

Key Stage 3 Assessment Tasks

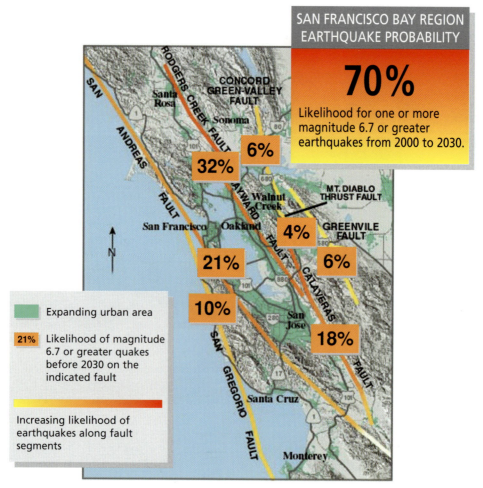

6 *A major earthquake is likely to hit San Francisco in the near future*

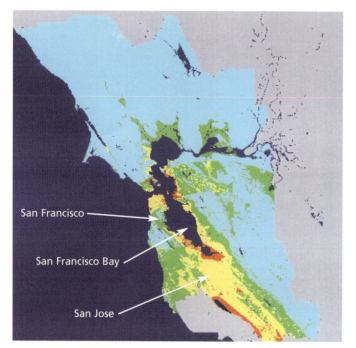

Modified Mercalli scale
- IX Violent
- VIII Very strong
- VII Strong
- VI Moderate
- V Light

The Modified Mercalli scale ranges from I to XII. It is used to describe earthquake intensity in relation to the amount of damage caused. A value of XII means 'total destruction'.

7 *The intensity of ground-shaking caused by an earthquake. Strong, or violent, shaking would cause much damage. The map shows the general level of risk, and does not relate to any particular event.*

Map Keys

OS 1:50,000 Landranger map series

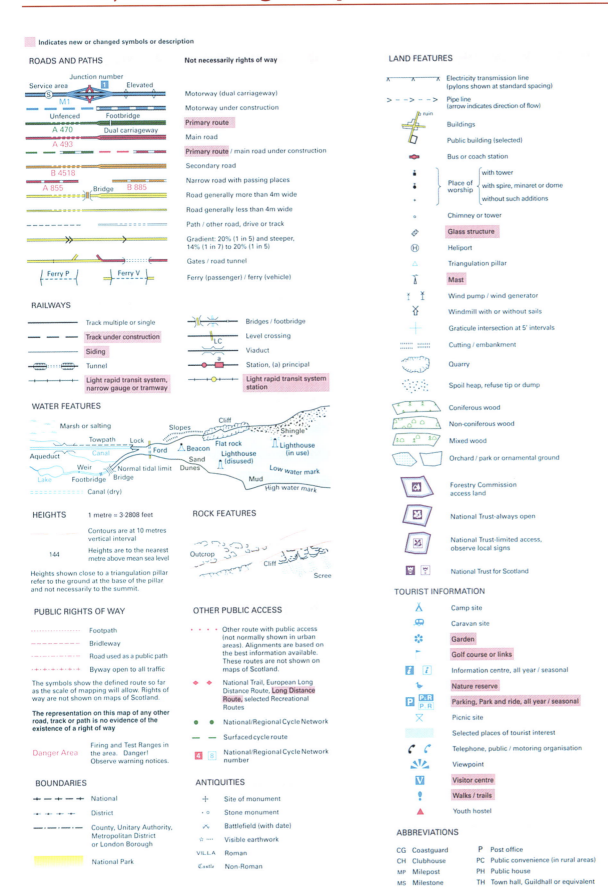

OS 1:25,000 Explorer map series

Land use maps

1960s
Key to map on page 36

SECOND LAND UTILISATION SURVEY OF BRITAIN

ARABLE LAND
- Cereals
- Ley legumes
- Roots
- Green fodder
- Industrial crops
- Fallow

MARKET GARDENING
- Field vegetables
- Mixed market gardening
- Nurseries
- Allotment gardens
- Flowers
- Soft fruit
- Hops

ORCHARDS
- With grass
- With arable land
- With market gardening

GRASSLAND

WOODLAND
- Deciduous
- Coniferous
- Mixed
- Coppice
- Coppice with standards
- Woodland scrub

WATER & MARSH
- Water
- Freshwater marsh
- Saltwater marsh

HEATH, MOORLAND ROUGH LAND

SETTLEMENT
- Commercial & Residential
- Caravan sites

UNVEGETATED

OPEN SPACE
- Tended but unproductive land

INDUSTRY
- Manufacturing
- Tips
- Public utilities

TRANSPORT
- Port areas, airfields, etc.
- Major roads
- Other metalled roads

DERELICT LAND

TYPES OF INDUSTRY
(overprinted on red)

3 Treatment of non-metalliferous mining products other than coal (glass, ceramics, cement, etc.)
4 Chemical and allied trades
6 Engineering, shipbuilding and electrical goods
8 Metal goods, not elsewhere specified
13 Food, drink and tobacco
14 Manufacture of wood and cork
16 Other manufacturing industries

HEATH, MOORLAND and ROUGH LAND
(overprinted on yellow)

C and/ or V Heather and/or Bilberry
S Wet Sphagnum
⌐ Scrub
U Gorse
G Grass Moor
F Festuca—Agrostis

1996

Key to map on page 37

KEY TO LAND USE SURVEY DATA

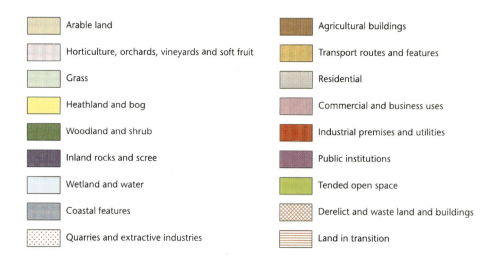

KEY TO ORDNANCE SURVEY DATA

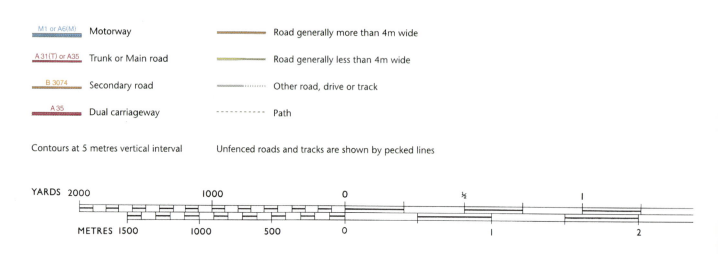

1:25,000 French survey maps

Key to maps on pages 62, 64 and 67.

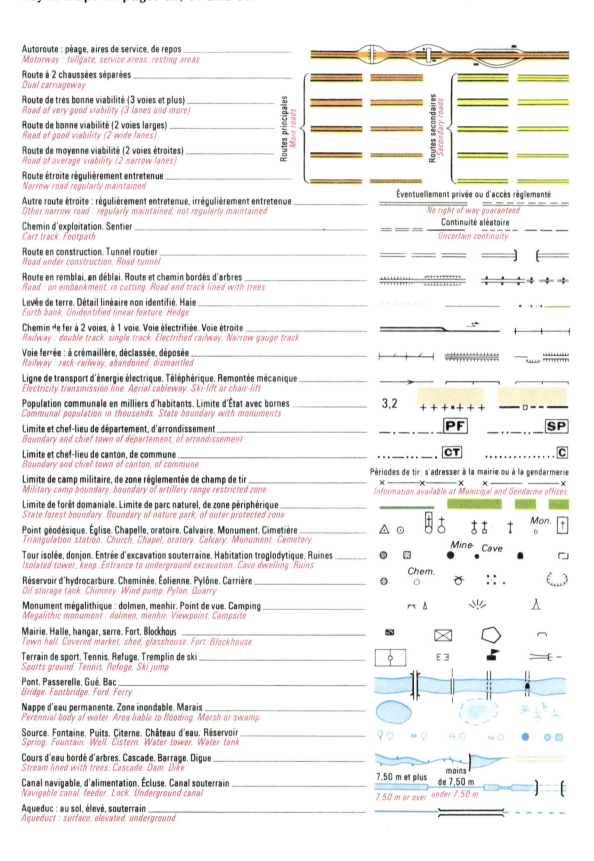

Phare. Feu. Bateau-feu. Épave
Lighthouse. Light. Lightship. Wreck

Sémaphore. Balise. Les courbes isobathes sont extraites des cartes du SHOM
Semaphore. Beacon. Depth contours are taken from the SHOM maps

Courbes de niveau, équidistance 5 m et 10 m. Dépression. Talus
Contours at 5 m and 10 m vertical interval. Depression. Slope

Bois de feuillus	Bois de conifères	Feuillus et conifères	Broussailles	Verger, plantation	Vigne	Rizière
Deciduous wood	*Coniferous wood*	*Deciduous and coniferous*	*Brushwood*	*Orchard, plantation*	*Vine*	*Rice field*

Équidistance des courbes 5 m et 10 m

1000 m   500 m   0   1 km

RENSEIGNEMENTS TOURISTIQUES
TOURIST INFORMATION

GR   autre sentier

Itinéraire balisé sur sentier (GR, autre sentier) (1), hors sentier (2)
Signposted route along footpath (GR, other) (1), out of footpath (2)

Itinéraire non balisé intéressant sur sentier
Interesting unsignposted route along footpath

Itinéraire de ski, de randonnée ou de raid
Cross-country or high mountain skiing route

Passage délicat
Hard part of hiking trail

Remontée mécanique en service en été
Ski-lift and chair-lift to be used in summer

Limite de zone réglementée
Boundary of restricted zone

Refuge ou gîte d'étape gardés, non gardés. Abri
Refuge hut or overnight stopping place with keeper, without keeper. Shelter

Camping. Centre équestre. Site d'escalade équipé. Aire de départ de vol libre
Campsite. Riding centre. Climbing site with facilities. Hang-gliding area

Aire de détente. Tennis. Golf
Leisure area. Tennis. Golf

Centre de ski de fond. Port de plaisance. Mouillage. Sports nautiques
Cross-country skiing centre. Yachting harbour. Anchorage. Sailing sports

Canoë-kayak (point de mise à l'eau). Piscine. Baignade
Canoeing (launching place). Swimming-pool. Bathing-place

Station classée
Resort with tourist interest

GEX

Ville d'art. Station thermale, verte, de sports d'hiver, balnéaire
City of artistic interest. Spa, openair, winter sports, seaside resort

Agglomération touristique, centre d'activité, site ou détail remarquables
Town of tourist interest, activity centre, notable site or building

FORET DE ROTHONNE

Édifice remarquable. Curiosité diverse. Informations tourisme
Notable monument. Divers place of interest. Tourist information centre

Gare ou point d'arrêt ouverts au trafic voyageurs
Station or stopping-place open to passenger traffic

Gare   Arrêt

Voie interdite aux véhicules à moteur. Aire de stationnement
Prohibited road for motor vehicles. Parking area

128

TEACHER'S NOTES

INVESTIGATIONS IN Applied Biology and Biotechnology

Peter Freeland

Hodder & Stoughton
LONDON SYDNEY AUCKLAND TORONTO

Acknowledgements

The author would like to express thanks to Mrs M Moore, laboratory assistant, for preparing materials for these investigations, and to his pupils, both past and present, for their part in obtaining results.

The author and publishers would like to thank the following for permission to use their photographs:

Teacher's notes
Victoria Wine Company p36; Metropolitan Police p47; Barnaby's Picture Library p52.

Pupil's sheets
Farmhouse Cheese Bureau p39; Tesco p41; Victoria Wine Company p58; Science Photo Library p76; Metropolitan Police p78; Barnaby's Picture Library p86; Science Photo Library p114; ICI Garden Products p116.

Data for Investigation 36, Fig 3 were supplied by Castlemead Publications, Ware, Herts. from charts constructed by Tanner and Whitehouse.

British Library Cataloguing in Publication Data
Freeland, P. W.
   Investigations in applied biology and biotechnology.
   1. Applied biology   2. Biotechnology
   I. Title
   630
   ISBN 0 340 50630 X

First published 1990

© 1990 Peter Freeland

This work is copyright.
Permission is given for copies to be made of the Pupil's sheets only, provided they are used exclusively within the Institution for which this work has been purchased. For reproduction for any other purpose, permission must first be obtained in writing from the publishers.
All rights reserved.

Typeset by Taurus Graphics, Abingdon, Oxon.
Printed in Great Britain for Hodder and Stoughton Educational, a division of Hodder and Stoughton Ltd, Mill Road, Dunton Green, Sevenoaks, Kent by St Edmundsbury Press, Bury St Edmunds, Suffolk.

Contents: Teacher's Notes

1 Enzyme isolation *5*
2 Enzymes in washing-powders *6*
3 The best temperature for a washing-powder *8*
4 Tenderising meat *11*
5 Peas and beans (pulses) *12*
6 An enzyme inhibitor in beans *13*
7 Protein precipitation *15*
8 Solvents in food manufacture *16*
9 Particle size in cereal products *17*
10 Physical properties of milk *19*
11 When does custard thicken? *21*
12 Making and using yogurt *22*
13 Cheese-making *23*
14 Cheese tasting *24*
15 Acid and sugar in citrus fruits *26*
16 Making sweeter sugar *27*
17 Energy content of foods *29*
18 'E' is for artificial colouring *30*
19 Natural colouring *31*
20 Research with extracted chloroplasts *32*
21 Bread-making *34*
22 Wine-making *35*
23 Malting barley *37*
24 Alcohol production *40*
25 Making a fermenter *41*

26 Cell counts *42*
27 Which antibiotic? *44*
28 Daily urine output *46*
29 Drinking and driving *47*
30 Variations in a human population *48*
31 Fingerprint analysis *49*
32 How exercise affects your heart *51*
33 Heart and breathing rates during exercise *52*
34 Lung capacity *54*
35 Measuring grip strength *55*
36 Obesity *56*
37 Successful seed germination *58*
38 New bulbs from old *60*
39 New plants from cuttings *61*
40 Rooting cuttings in a gel *63*
41 Micropropagation *64*
42 Pollination *66*
43 Longer life for cut flowers *67*
44 Dispersing weed seeds *68*
45 Trapping and killing slugs *70*
46 Soil conditioning *71*
47 Composting *73*
48 Making a biogas generator *74*
49 The effects of acid rain *76*
50 Water purification *77*

Contents: Pupil's Sheets

These follow page 80 of the Teacher's notes.

Preface

These investigations have been written for pupils in the third year upwards, particularly those working towards GCSE and Standard Grade examinations in biology and science. In the past, biologists have concentrated on the structure of organisms (morphology/ anatomy), the functions of their different organs (physiology) and their relationships with organisms and their environment (ecology). Today, this outlook has broadened in the direction of applied biology, including health education and biotechnology. Applied biology is the application of biological knowledge to specific problems, such as the control of pests or the improvement of soils. Biotechnology is more difficult to define. Primarily, it is the commercial use of living organisms, and some of their products, for human purposes. It also involves engineering skills in the construction of manufacturing plants suitable for the large-scale production of products such as proteins, enzymes, antibiotics, foods and industrial solvents.

The pack contains 50 practical investigations, designed to introduce some aspects of applied biology and biotechnology. The topics are wide ranging. Coverage includes enzymes, food manufacture and fermentation technology, together with health education, plant propagation, pest control, waste disposal and soil conditioning. Most of the investigations involve carrying out a number of tasks of varying difficulty. Each investigation is followed by questions, with an accompanying mark scheme designed to test understanding and to assess specific scientific skills. All of the investigations end with a section called 'Taking it further'. This contains one or more suggestions for follow-up work, which might be suitable for individual projects.

By working through these investigations pupils should gain a better understanding of the ways in which biologists maintain, and try to improve, the quality of human life.

Peter Freeland 1989

1 Enzyme isolation

Preparation: *Part 1 – 5 mins*
Part 2 – 20–30 mins (after 2–3 days)
Investigation: *30–40 mins*

Preparation

MATERIALS

Part 1

- 100 g pea or bean seeds
- 500 cm^3 beaker

Part 2

- soaked pea or bean seeds in 500 cm^3 beaker
- starch-agar plate
- 100 cm^3 beaker
- 100 cm^3 measuring cylinder
- 5 test tubes fitted with rubber bungs
- 5 1 cm^3 plastic syringes
- filter funnel
- no. 6 cork borer
- kitchen mixer
- waterbath maintained at 30°C
- incubator maintained at 25°C
- filter papers
- adhesive tape
- glass-marking pen
- access to refrigerator

Investigation

MATERIALS

- starch-agar plate containing seed extracts
- iodine solution
- ruler, graduated in mm

Instructions and notes

1 Any species of pea or bean seed may be used as a source of amylase.

2 A method for preparing starch-agar plates is given on page 00.

3 When filtering the slurry obtained by grinding germinating seeds with water, use a fine-grained filter paper for filtration. Whatman grade no. 540, for 8 μm particle retention, is suitable.

4 Bench iodine solution may be used. Alternatively, prepare a solution of iodine by dissolving 4.0 g potassium iodide and 0.2 g iodine in 100 cm^3 water.

Teacher's notes   5

Specimen answers

1 (a)

Hole no.	Diameter of clear zone (cm)	Area of clear zone (cm²)
1	1.1	0.95
2	1.3	1.33
3	1.9	2.8
4	2.6	5.3
5	3.6	10.2

(5)

(b) (i) yes, (ii) yes. (2)

2 (a) $\dfrac{100}{75} \times 1.0 = 1.33$ **(b)** $\dfrac{100}{50} \times 1.0 = 2.0$

(c) $\dfrac{100}{25} \times 1.0 = 4.0$ (3)

(d)

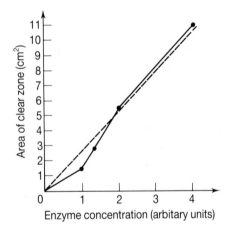

Fig 1 (4)

(e) There is a direct relationship between the concentration of the enzyme and the area of clear zone. (2)

3 Extracellular amylase activity is indicated by the clear zone around hole 1 (0.95 cm²). The clear zone around hole 2 indicates extracellular + intracellular activity (1.33 cm²).

$$\text{percentage extracellular enzyme} = \dfrac{0.95}{1.33} \times 100$$
$$= 71.4\%$$

$$\text{percentage intracellular activity} = \dfrac{1.33 - 0.95}{1.33} \times 100$$
$$= 28.6\% \quad (4)$$

2 Enzymes in washing-powders

Preparation (day 1): *20–30 mins*
Investigation (day 2): *30–40 mins*

Preparation

MATERIALS

- 5 cm³ 1%^(w/v) solution of powder A in a labelled container
- 5 cm³ 1%^(w/v) solution of powder A₁ in a labelled container

6 Teacher's notes

- 5 cm^3 1%$^{w/v}$ solution of powder B in a labelled container
- starch-agar plate
- milk-agar plate
- mayonnaise-agar plate
- 1 cm^3 plastic syringe
- no. 6 cork borer
- adhesive tape
- glass-marking pen
- incubator maintained at 25°C

Investigation

MATERIALS

- incubated starch-agar plate
- incubated milk-agar plate
- incubated mayonnaise-agar plate
- iodine solution
- ruler, graduated in mm

Instructions and notes

1 This investigation uses two different brands of biological washing-powder, A and B. The powder labelled A$_1$ is the same product as A, but has been stored for at least 6 months. Alternatively, powder A$_1$ may be produced by ageing powder A artificially. Heat some of powder A in an incubator at 50°C for 10−20 minutes.

2 The agar plates should be prepared from a refined powder, such as bacteriological agar. Plates should be poured to a depth of 0.5−0.75 cm.

Starch agar:
Mix 2 g soluble starch with cold water to form a slurry. Add this to 100 cm^3 boiling water. Mix 2 g agar powder with cold water. Add this to the boiling starch solution, stirring to prevent localised heating.

Milk agar:
Mix 2 g Marvel milk powder with 100 cm^3 cold water. Heat the mixture until it boils. Add 2 g agar powder, mixed to a slurry with cold water. Stir continuously until the agar has dissolved.

Mayonnaise agar:
Mix a heaped teaspoon of mayonnaise with 100 cm^3 cold water. Proceed as in the preparation of milk agar.

3 Alternative agars for detecting lipase activity may be prepared with salad cream (1−2% $^{w/v}$) and egg yolk (0.5−1.0% $^{w/v}$). Some suppliers of biological materials (3, 4) offer gelatinised tributyrin agar, which can be melted and used for the same purpose.

Teacher's notes   7

Specimen answers

1

Agar plate	Solution	Diameter of clear zone (cm)	Area of clear zone (cm²)
starch	A	1.6	2.01
	A_1	1.2	1.13
	B	1.4	1.52
milk	A	1.4	1.52
	A_1	0.8	0.50
	B	1.2	1.13
mayonnaise	A	0.0	0.0
	A_1	0.0	0.0
	B	0.0	0.0

(9)

2 A = amylase, protease
A_1 = amylase, protease
B = amylase, protease   (6)

3 Powder A was the most active. It has more amylase and protease activity.   (2)

4 (a) As a result of storage, solution A_1 showed less amylase and protease activity than solution A.
(b) The percentage decrease in amylase activity is $\dfrac{2.01 - 1.13}{2.01} \times 100 = 40\%$ (3)

---

3 The best temperature for a washing-powder

Preparation: *50–60 mins*
Investigation: *30–40 mins*

Preparation

MATERIALS

- 5 cm³ 1%$^{w/v}$ trypsin solution in a 5 cm³ plastic syringe
- 5 cm³ 1%$^{w/v}$ biological washing-powder solution in a 5 cm³ plastic syringe
- 2 milk-agar plates
- 250 cm³ beaker
- thermometer
- no. 6 cork borer

8   Teacher's notes

- adhesive tape
- Bunsen burner, tripod and gauze
- paper towel
- incubator maintained at 25°C
- glass-marking pen
- eye protection

Investigation

MATERIALS

- 2 incubated milk-agar plates
- ruler, graduated in mm

Instructions and notes

1 Pupils require a 1%$^{w/v}$ solution of (a) trypsin and (b) a biological washing-powder. These solutions should be prepared immediately before they are required and supplied in 5 cm^3 plastic syringes.

2 A method for preparing milk-agar is given on page 7.

3 **Warnings should be given about the need for care when holding hot syringes, which should be held only by the plunger and flange at the top of the barrel.**

Specimen answers

1

Hole no.	Temperature (°C)	Diameter of clear zone (cm) Trypsin	Diameter of clear zone (cm) Washing-powder	Area of clear zone (cm^2) Trypsin	Area of clear zone (cm^2) Washing-powder
1	30	1.8	1.9	2.54	2.83
2	40	1.8	1.9	2.54	2.83
3	50	1.7	1.8	2.26	2.54
4	60	1.4	1.8	1.53	2.54
5	70	1.0	1.8	0.78	2.54
6	80	0.0	1.2	0.0	1.13
7	90	0.0	1.0	0.0	0.78

(14)

2

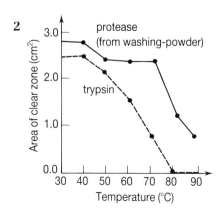

(5)

3 **(a)** Percentage decrease in trypsin acvitity at $60°C = \dfrac{1.01}{2.54} \times 100 = 39.7\%$

(b) Percentage decrease in protease (washing-powder) activity at 60°C
$= \dfrac{0.29}{2.83} \times 100 = 10.2\%$ (4)

4 The investigation shows that the protease in washing-powder has a higher optimal temperature than trypsin. Furthermore, this protease retains more of its activity at temperatures in the range 40–70°C, while trypsin undergoes rapid denaturation at these temperatures. This relative resistance to denaturation by heat makes the protease particularly suitable for use in washing-powders. (4)

5 1. The temperature inside the syringes is probably lower than that in the surrounding water.
2. Exact temperatures of 40, 50, 60°C etc. were difficult to maintain for a period of 3 minutes.
3. At least two milk-agar plates should have been used for each enzyme solution so that results could have been duplicated. (3)

Tenderising meat

Preparation: 20–30 mins
Investigation: 30–40 mins

Preparation

MATERIALS

- steak (400–500 g)
- 6 bottles, each containing 100 cm³ of a buffer solution (e.g. pH 4.0, 6.8, 7.0, 8.0, 9.0 and 9.2)
- 6 100 cm³ beakers
- forceps
- scalpel
- glass-marking pen
- Clingfilm
- top-pan balance

Investigation

MATERIALS

- 6 100 cm³ beakers containing cubed steak in buffer solutions
- forceps
- paper tissues
- top-pan balance

Instructions and notes

1 The meat should be soaked for 12–24 hours before the effects of pH are investigated. It is suggested that each class should carry out a single investigation, and the results made available to each pupil.

2 It is best to use pH tablets that span the complete pH range. Even so, the investigation can be carried out using any six different tablets, depending on availability.

3 **At the end of the investigation pieces of cubed steak should be incinerated.**

Specimen answers

1

pH of buffer	Mass of meat (g) Before soaking	Mass of meat (g) After soaking	Increase (+)/ decrease (−) (g)	% increase/ decrease
4.0	21.3	25.4	+1.1	+5.16
6.8	25.6	26.1	+0.5	+1.9
7.0	27.5	27.6	+0.1	+0.4
8.0	29.0	28.2	−0.8	−2.7
9.0	27.7	26.8	−0.9	−3.3
9.2	24.9	23.9	−1.0	−4.0

2

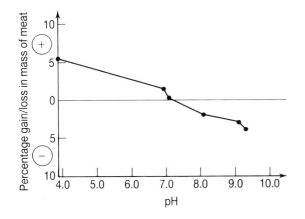

(4)

3 (a) When soaked in an acid–neutral solution, meat absorbs water. In an alkaline solution, however, water is lost. (It is this loss of water that leads to toughness during cooking.) (2)

(b) Only in an acid solution does the meat retain its red colour. It becomes light brown/white in neutral and alkaline solutions. (2)

4 *Criticisms*
1. The choice of pH buffers did not completely span the pH range.
2. Each beaker contained a different amount of meat.
3. Individual pieces of meat were of different sizes, with different surface areas exposed to the surrounding solutions.

Improvements
1. Use pH buffers 3.0, 4.0, 5.0 etc., to 14.0.
2. Put the same amount of meat into each beaker.
3. Use uniformly sized cubes or slices so that each piece of meat has the same surface area for water absorption/loss. (6)

$\mathbf{5}$ Peas and beans (pulses)

Preparation: *10–15 mins*
Investigation: *60–80 mins*

Preparation

MATERIALS

- 6 different types of pea/bean
- 6 boiling tubes in a rack
- 65 cm^3 0.01 M NaOH solution, coloured pink with phenolphthalein (indicator solution)
- 10 cm^3 plastic syringe
- ruler, graduated in mm
- glass-marking pen

Investigation

MATERIALS

- 6 boiling tubes containing soaked seeds
- 6 (or more) different types of dry seeds (numbered)
- 6 Clinistix reagent strips
- dropping bottle containing 10-vol H$_2$O$_2$ solution
- 6 text tubes in a rack
- white tile
- scalpel
- ruler, graduated in mm
- glass-marking pen
- plastic gloves
- eye protection

Instructions and notes

1 The pH indicator is prepared by mixing 100 cm^3 0.1 M NaOH solution, 900 cm^3 distilled water and 10 cm^3 bench phenolphthalein indicator. Each pupil will need 65 cm^3 of the indicator.

2 If the key is used to identify peas/beans, select six or more types, and number each seed with a glass-marking pen. In trials, each pupil was supplied with six types of seed, numbered as follows:
pea (1); dwarf French bean (2); broad bean (3); soya bean (4); butter (lima) bean (5); black-eyed bean (6).

3 Hydrogen peroxide solution is acid and can cause skin burns. Pupils should wear plastic gloves and eye protection when handling this reagent.

Specimen answers

1

Test	Seeds					
	A	B	C	D	E	F
indicator colourless (C)/pink (P)	C	C	C	C	C	C
glucose present (P)/absent (A)	P	P	P	P	P	P
oxygen present (P)/absent (A)	P	P	P	P	P	P

(18)

2 All the seeds tested were living. They all released CO_2 and glucose into their surroundings, and produced bubbles of oxygen when hydrogen peroxide was added. (2)

3

No. pea/bean	Sequence of key numbers used in identification	Name
1	1, 2	pea
2	1, 3, 4, 5, 7, 9	dwarf French bean
3	1, 3, 4	broad bean
4	1, 3	soya bean
5	1, 3, 4, 5, 6	butter (lima) bean
6	1, 3, 4, 5, 7, 8	black-eyed bean

(5)

6 An enzyme inhibitor in beans

Preparation: day 1 – 5 mins
day 2 – 30 mins
Investigation: 30–50 mins

MATERIALS

Day 1

- 200 g red kidney beans
- 250 cm³ beaker
- household bleach

Teacher's notes 13

Day 2

- soaked kidney beans
- milk-agar plate
- 5 cm^3 1%$^{w/v}$ trypsin solution
- 3 100 cm^3 beakers
- 4 1 cm^3 plastic syringes
- no. 6 cork borer ·
- glass-marking pen
- adhesive tape
- Bunsen burner
- kitchen mixer
- incubator maintained at 25°C
- eye protection

Investigation

MATERIALS

- petri dish containing trypsin and bean extracts
- ruler, graduated in mm

Instructions and notes

1 If red kidney beans are not available, black-eyed beans, dwarf French beans, haricot beans or runner beans may be used.

2 A method for preparing milk-agar is described on page 7.

Specimen answers

1

Hole no.	Diameter of clear zone (cm)	Area of clear zone (cm^2)
1	1.6	2.01
2	1.6	2.01
3	0.0	0.0
4	1.6	2.01

(6)

2 Uninhibited trypsin digests milk from an area of about 2.01 cm^2. The bean water does not contain an inhibitor, as the trypsin digests milk protein from a similar area around hole 2. Raw bean pulp contains a powerful trypsin inhibitor. This inhibitor must be present in the cells, as it is present only following grinding. Boiled bean pulp has no effect on trypsin activity. This suggests that boiling the beans destroys the inhibitor, and that the inhibitor is probably a protein. (5)

3 (a) If bleach, or some other antiseptic, had not been added to the beans while they were soaking, bacteria and fungi might have grown in the water. Protein-digesting enzymes from the bacteria and fungi could have given false readings when the bean water was put into the plate of milk agar. (2)

(b) If water had not been added to trypsin in hole 1, this trypsin would be of double strength. Owing to its smaller volume it would probably have diffused into the agar at a slower rate than the trypsin in the other holes. (2)

14   Teacher's notes

7 Protein precipitation

Investigation: *30–40 mins*

Investigation

MATERIALS

- 20 cm^3 soluble casein[1] solution in a labelled beaker
- 20 cm^3 dilute (2 M) hydrochloric acid in a labelled beaker
- 20 cm^3 distilled water in a labelled beaker
- 5 numbered test tubes in a rack, each with a different precipitating agent
- 6 test tubes
- 2 5 cm^3 plastic syringes
- glass-marking pen

Instructions and notes

1 The casein solution contains 2.5 g soluble casein/100 cm^3 water. Slight opacity will not affect the result.

2 Each pupil should be supplied with a rack containing five numbered test tubes, each containing 2 cm^3 of the following precipitating agents:

tube 1: saturated sodium chloride solution
tube 2: dilute (2 M) sulphuric acid
tube 3: ethanol
tube 4: saturated ammonium sulphate solution
tube 5: dilute (2 M) sodium hydroxide solution

Saturated sodium chloride solution contains approximately 30 g/ 100 cm^3 water, while saturated ammonium sulphate solution contains approximately 50 g/100 cm^3 water.

3 Label the beakers containing casein solution, hydrochloric acid and distilled water.

Specimen answers

1 (a)

Tube no.	Precipitating agent	Amount of precipitate
1	sodium chloride	√
2	sulphuric acid	√√√
3	ethanol	×
4	ammonium sulphate	√√
5	sodium hydroxide	×

(5)

(b) sulphuric acid   (1)

Teacher's notes   15

2 (a)

Tube no.	Vol. water (cm³)	Vol. HCl (cm³)	Concentration of HCl	Amount of precipitate
1	0.0	5.0	2.0 M	✓✓✓
2	1.0	4.0	1.6	✓✓
3	2.0	3.0	1.2	✓
4	3.0	2.0	0.8	✓
5	4.0	1.0	0.4	✗

(8)

(b) The lowest concentration of HCl that will precipitate casein is 0.8 M. (1)
(c) As the precipitating agent (HCl) may have to be removed from the product (casein), it is important to use the lowest possible concentration of the precipitating agent. Using precipitating agents at their lowest effective concentration also saves money. (2)
(d) Extraction (by filtration), removal of acid, drying. (3)

Solvents in food manufacture

Investigation: *30–45 mins*

Investigation

MATERIALS

- 3–5 g porridge oats
- 2 Clinistix reagent strips[3,4]
- 2 Albustix reagent strips[3,4]
- 4 test tubes fitted with corks or rubber bungs
- test tube rack
- 20 cm³ solvent A
- 20 cm³ solvent B
- 20 cm³ solvent C
- 20 cm³ solvent D
- iodine solution
- 1 cm³ plastic syringe
- 10 cm³ plastic syringe
- ruler, graduated in mm
- scissors
- glass-marking pen

Instructions and notes

1 Use a finely milled oatmeal, such as Ready-Brek. As this product contains little or no glucose, add powdered D-glucose at the rate of about $\frac{1}{4}$ teaspoon/100 g. Mix thoroughly.

2 The following solvents are required:

solvent A = tap water
solvent B = 20%$^{w/v}$ sodium chloride solution
solvent C = 20%$^{w/v}$ alcohol (ethanol) solution
solvent D = 0.1 M hydrochloric acid

Each pupil should be provided with about 20 cm^3 of each solvent in a labelled container. **Pupils should be warned that solvent D is acid.**

Specimen answers

1

Tube	Length of oatmeal column (cm)	
	Top	Bottom
A	1.0	1.2
B	1.6	0.7
C	1.4	0.6
D	0.3	1.8

(6)

2 (a) tube B **(b)** tube D **(c)** tube A (3)

3 A, C, D, B (4)

4 C, A. No protein extracted by B or D. (4)

5 D, A, B, C (4)

6 A suitable mass of glucose (e.g. 0.5 g) should be added to 10 cm^3 solvent. Solutions should be tested with Clinistix.
A suitable mass of egg albumen, or another protein, should be added to 10 cm^3 solvent. Solutions should be tested with Albustix. (Note: both glucose and proteins have different solubilities in the four solvents. This accounts for the differences.) (4)

9 Particle size in cereal products

Investigation: *40–60 mins*

Investigation

MATERIALS

- 50 g porridge oats
- 50 g wholemeal flour

Teacher's notes 17

- 50 g white flour
- 100 cm³ beaker (container for weighing samples)
- top-pan balance
- nest of sieves, with lid and receiver[3]

Instructions and notes

1 A single sieving should provide data for all members of a class.

2 Several biological suppliers offer nests of sieves, containing six or more sieving units, each with a different nominal aperture. The size of these apertures is always specified by the manufacturer.

3 Sieving should be carried out by placing the sieve, with the receiver beneath and lid on top, on the bench. The sieve should then be slid swiftly and sharply from side to side until sifting is complete. Note that the use of any kind of stirring rod produces false results, as large particles are broken down by friction.

Specimen answers

1 (5)

	Product		
Aperture (mm)	porridge oats	wholemeal flour	white flour
1.7	23.9	0.1	0.0
0.85	17.1	2.9	0.0
0.25	7.1	28.5	0.0
0.18	1.1	12.4	0.1
< 0.18	0.8	4.7	48.3

2

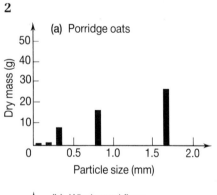

3 (a) porridge oats (b) wholemeal flour
   (c) white flour (3)

4 (a) porridge oats (b) white flour (2)

5 1. Some fine particles became trapped in the mesh so that the combined masses of sieved material did not add up to 50 g.
2. Shaking material in a sieve may break up some of the particles. Hence the results may underestimate the mass of larger particles, but overestimate the mass of smaller ones. (2)

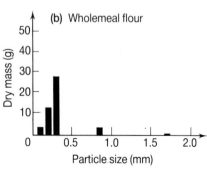

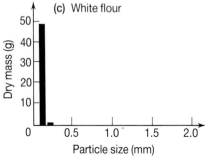

(8)

18 Teacher's notes

10 Physical properties of milk

Investigation: *50–70 mins*

Investigation

MATERIALS

- 70 cm³ full cream milk
- 70 cm³ skimmed milk
- 2 test tubes fitted with rubber bungs
- 2 100 cm³ beakers, graduated in 25 cm³ units
- 50 cm³ measuring cylinder
- 10 cm³ plastic syringe
- filter funnel
- 2 thermometers
- retort stand, boss and clamp
- scalpel
- ruler, graduated in mm
- glass-marking pen
- matchstick
- waterbath maintained at 90°C
- stop-clock, or watch with a second hand

Instructions and notes

1 Full cream milk and skimmed milk are available from most large supermarkets.

2 This is a fairly demanding investigation, requiring accurate measurements to be made at regular intervals.

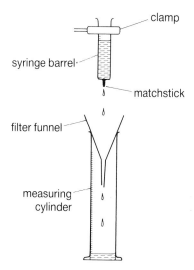

Specimen answers

1 **(a)** (i) 1.1 cm   (ii) 0.0 cm   (2)

 (b) (i) $\frac{1.1}{5.0} \times 100 = 22\%$ (ii) 0.0   (2)

Teacher's notes 19

2

	Vol. milk (cm³)				
	Full cream		Skimmed	
Time (seconds)	In syringe	Drained out	In syringe	Drained out
0	10.0	0.0	10.0	0.0
30	6.9	3.1	6.4	3.6
60	5.0	5.0	4.7	5.3
90	3.1	6.9	2.7	7.3
120	1.7	8.3	1.2	8.8
150	0.9	9.1	0.3	9.7
180	0.6	9.4	0.0	10.0
210	0.3	9.7		
240	0.0	10.0		

(6)

3 (a)

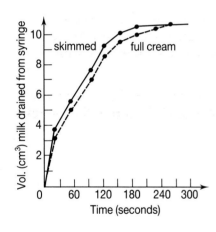

Fig 2 (5)

(b) The more cream present, the slower the flow rate of the milk. (2)

4 (a)

Time (minutes)		0 2 4 6 8 10 12 14 16 18 20 22 24 26 28 30 32 34 36
Temperature (°C)	Full cream	90 82 74 63 58 53 50 47 44 41 40 38 36 34 33 32 31 30
	skimmed	90 82 74 65 58 54 50 47 44 42 40 38 36 34 33 31 29

(4)

(b) Both types of milk lost heat at about the same rate. However, a skin had formed on the full cream milk at 60°C. This skin probably retained some heat so that the full cream milk cooled more slowly over the temperature range 30–40°C. (4)

11 When does custard thicken?

Investigation: *30–40 mins*

Investigation

MATERIALS

- 2 g custard powder
- 2 100 cm^3 beakers
- 5 cm^3 plastic syringe
- thermometer
- tripod and gauze
- Bunsen burner
- stop-clock, or watch with a second hand
- eye protection

Instructions and notes

1 You must emphasise that the custard should be heated slowly, with frequent withdrawal of the heat source. Again, emphasise that the mixture should not be heated to boiling point, otherwise it will boil over.

2 A safe method for withdrawing hot samples, by holding both the handle of the syringe and the flange at the top of the barrel, should be demonstrated.

Specimen answers

1

Temperature (°C)	Time for drainage of syringe (s)
room	5.0
25	5.0
35	5.0
45	5.0
55	5.5
65	6.0
75	10.0
85	13.0
95	18.0

(8)

2

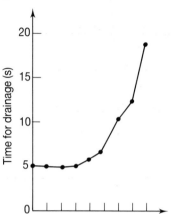

3 Gelatinisation occurred over the temperature range 65–95°C. (2)

(5)

12 Making and using yogurt

Investigation: *30–40 mins*

Investigation

MATERIALS

- 200 cm³ milk in a plastic beaker
- yogurt 'starter' culture or teaspoon of live yogurt
- 250 cm³ beaker
- 1 dm³ beaker
- thermometer
- teaspoon
- cloth (for handling hot beakers)
- Bunsen burner, tripod and gauze
- incubator maintained at 43°C
- access to refrigerator
- Clingfilm
- eye protection

Instructions and notes

1 The use of a freeze-dried or liquid 'starter' culture is recommended. Live yogurt bought in a shop can provide a culture of the necessary bacteria, but there is no guarantee that sufficient lactic acid will be produced. Sometimes it can work well, while at other times nothing happens.

Specimen answers

1 (a) Boiling water kills any bacteria present in the beaker. (1)

(b) These bacteria could have produced undesirable products/flavours. (1)

22 Teacher's notes

2 Bacteria in the milk are killed by heating to 82°C. (1)

3 Milk tends to boil over when heated to 100°C. Also, milk that has been boiled tends to form a skin as it cools. (2)

4 Bacteria in the 'starter' culture would be killed at temperatures above 43°C. (1)

5 A longer period of fermentation would make the yogurt more acid and less palatable. (1)

6 1. The live culture did not contain any living bacteria.
2. The 'starter' culture was added at too high a temperature (kills bacteria).
3. The mixture had not been allowed to ferment for long enough. (2)

7 The yogurt may be contaminated with foreign bacteria, including species that could cause disease. (1)

13 Cheese-making

Investigation: *Part 1 – 30–50 mins*
Part 2 – 10–15 mins
Part 3 – 20–40 mins

Investigation

MATERIALS

Part 1

- 1 pint milk
- cheese 'starter' culture[3,4]
- 10 cm^3 rennet solution[5]
- large freezer bag
- large saucepan
- thermometer
- string
- access to cooker

Part 2

- coagulated milk (in freezer bag)
- 1 dm^3 beaker
- retort stand, boss and clamp
- pin

Part 3

- cheese curd (in freezer bag)
- 2 g salt
- muslin, approximately 15 × 15 cm
- aluminium foil, approximately 15 × 15 cm
- cardboard cylinder, approximately 8 cm long × 6 cm wide
- teaspoon
- access to boiling water (e.g. in kettle)

Teacher's notes   23

Instructions and notes

1 The rennet solution contains 1 g rennin (or 1 cm³ rennet concentrate) dissolved in 100 cm³ water.

2 Commercial freeze-dried or liquid 'starter' cultures may be used, added according to the manufacturer's instructions. Alternatively, 'starter' cultures may be prepared from mature cheeses as follows:

Boil 250 cm³ milk. Cool to 35°C and stir in 10–30 g finely grated cheese, cut from the region immediately beneath the rind. Keep the starter in a refrigerator until it is needed.

3 A cardboard cylinder, to serve as a mould for salted cheese curd, can be cut from cardboard at the centre of a roll of kitchen paper or aluminium foil.

Specimen answers

1 (a) lactose **(b)** lactic acid bacteria
  (c) lactic acid (3)

2 rennin (1)

3 (a) 'Unripened' cheeses (e.g. cottage, cream) consist of raw coagulated curd. They are eaten soon after production, without any further treatment. 'Ripened' cheeses (e.g. Cheddar, Danish blue) are made by treating raw cheese curd with specific strains of bacteria or fungi. These cheeses are eaten only after they have matured. (2)

(b) Hard cheese is made from compressed curd, internally ripened by bacteria or fungi. Soft cheese is made from uncompressed or compressed curd, ripened by bacteria or fungi that grow into the cheese from the outside. (2)

4 The main aims of the 'ripening' process are to enhance flavour and change the rubbery texture of raw cheese curd into something that is more creamy, crumbly and palatable. (2)

14 Cheese tasting

Investigation: *15–25 mins*

Investigation

MATERIALS

- 5 or more 'cheese-tasters'
- New Zealand Cheddar, labelled A
- farmhouse Cheddar, labelled B
- mature Cheddar, labelled C

24   Teacher's notes

Instructions and notes

1 This is a fairly demanding exercise, requiring the presentation and interpretation of data.

2 Care should be taken to avoid food being eaten in the laboratory. Pupils should be asked to carry out the tasting at home. They could, for example, be supplied with small pieces of cheese, approximately $2 \times 2 \times 2$ cm, in labelled plastic bags.

3 If pupils are to analyse results from a group of five or more pupils, each pupil should be asked to prepare five or more copies of their responses, or write them on the blackboard.

4 This type of investigation can be applied to a wide range of foods and soft drinks. If meaningful results are to be obtained however, it may be necessary to draw up a list of words that describe favourable, neutral and unfavourable responses to a product.

Specimen answers

1

Pupil no.	Rank order of cheeses		
	1st	**2nd**	**3rd**
1	A	C	B
2	C	A	B
3	A	C	B
4	A	C	B
5	A	C	B

(4)

2 1st-rated cheese: 3 points
2nd-rated cheese: 2 points
3rd-rated cheese: 1 point

Pupil no.	Cheese		
	A	**B**	**C**
1	3	1	2
2	2	1	3
3	3	1	2
4	3	1	2
5	3	1	2
total	14	5	11

(5)

Teacher's notes 25

Description	Cheese		
	A	B	C
peppery	0	3	0
salty	0	1	0
mild	5	0	1
creamy	2	0	1
musty	0	3	0
nutty	0	0	4
gritty	0	0	0
sickly	0	3	0
smooth	1	0	4
acid	2	0	0

(4)

4 Instruct tasters to taste each cheese in turn, then select one word to describe (i) its best feature and (ii) its worst feature. Record choices. Draw two tables, similar to that above, one listing features that tasters liked best, and the other listing features they liked least.　(4)

5 The investigation was confined to a small sample, made up of individuals from the same age group. The sample size should be increased to at least 100. People of all ages should act as tasters.　(3)

15 Acid and sugar in citrus fruits

Investigation: *40–60 mins*

Investigation

MATERIALS

- orange
- lemon
- 2 Clinitest tablets
- Clinitest colour chart
- 35 cm^3 0.1 M NaOH solution
- 25 cm^3 1%$^{w/v}$ citric acid solution
- phenolphthalein indicator in a dropping bottle
- 2 flat-bottomed tubes or test tubes in a rack
- 2 250 cm^3 beakers
- 100 cm^3 beaker
- 100 cm^3 measuring cylinder

- 10 cm³ plastic syringe
- 5 cm³ plastic syringe
- 1 cm³ plastic syringe
- scalpel
- forceps
- glass-marking pen
- top-pan balance

Instructions and notes

1 Small oranges and lemons are suitable for this investigation. Alternatively, each pupil could be supplied with half an orange and lemon.

2 Clinitest tablets should not be handled, as they contain sodium hydroxide. When water is added to a tablet, the mixture boils so that the tube may become hot. Pupils should be warned of this.

Specimen answers

1 **(a)** 146 g   **(b)** 102 g   (2)

2 **(a)** 66 cm³   **(b)** 42 cm³   (2)

3 **(a)** $\dfrac{66}{146} \times 100 = 45.2\%$

(b) $\dfrac{42}{102} \times 100 = 41$ (4)

4 5.4 cm³   (1)

5 **(a)** 4.1 cm³   (1)

(b) $\dfrac{5.4}{4.1} \times 1.0 = 1.31$ g/100 cm³   (1)

6 **(a)** 1,1 cm³   (1)

(b) $\dfrac{5.4}{1.1} \times 1.0 = 4.9$ g/100 cm³   (1)

(c) It is assumed that all the acid in the juice is citric acid.   (1)

7 **(a)** 1% +   **(b)** 2% +   (2)

8 Compared with orange juice, lemon juice contains more citric acid and reducing sugars. Even so, high concentrations of citric acid in lemon juice completely mask the sweet taste of the reducing sugars. Orange juice does not contain enough citric acid to mask the sweet taste of the reducing sugars it contains.   (6)

16 Making sweeter sugar

Investigation: *50–80 mins*

Investigation

MATERIALS

- 5 cm³ invertase concentrate[1]
- 30 cm³ 3%$^{w/v}$ sodium alginate solution in a beaker
- 50 cm³ 3%$^{w/v}$ calcium chloride solution in a beaker

Teacher's notes   27

- 150 cm³ 2%^{w/v} sucrose solution
- 5 Clinitest tablets
- Clinitest colour chart
- 250 cm³ beaker
- 5 specimen tubes, approximately 8 × 1.5 cm
- 10 cm³ plastic syringe
- retort stand, boss and clamp
- kitchen sieve
- freezer bag
- test tube holder
- forceps
- glass rod
- pin
- stop-clock, or watch with a second hand

Instructions and notes

1 Pupils should be shown the technique for producing uniformly sized beads of sodium alginate. The enzyme/alginate mixture should be released slowly, and at a steady rate, from the syringe.

2 Use freezer bags, rather than plastic bags, to hold the sucrose solution. Thin plastic bags, such as those sold to hold food, may split or leak when filled with a solution.

Specimen answers

1 (a)

Time (minutes)	0	5	20	15	20	25
% glucose	0	0.5	0.75	1.0	2.0	2.0

(4)

(b)

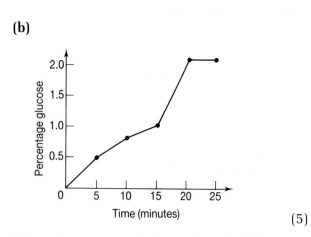

(5)

2 The solution must be heated until all the water has evaporated away, leaving only solid glucose. (2)

3 It would be necessary to (i) know when the reaction   sucrose → glucose + fructose   was complete, and (ii) remove fructose from the end product. (4)

4

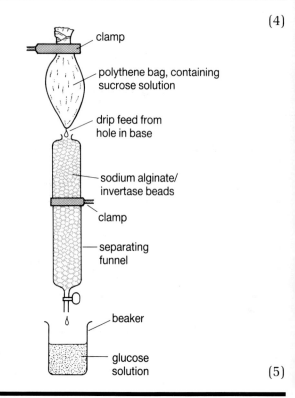

(5)

28   Teacher's notes

17 Energy content of foods

Investigation: *40–50 mins*

Investigations

MATERIALS

- $\frac{1}{2}$ dry-roasted peanut, labelled A
- $\frac{1}{2}$ oil-roasted peanut, labelled B
- piece of cream cracker biscuit, labelled C
- piece of digestive biscuit, labelled D
- 4 test tubes
- test tube rack
- 10 cm³ plastic syringe
- thermometer
- mounted needle
- glass-marking pen
- retort stand, boss and clamp
- Bunsen burner
- eye protection

Instructions and notes

1 A small packet of (i) dry-roasted and (ii) oil-roasted peanuts will be enough for a class of 20 or more pupils. Each pupil should be given $\frac{1}{2}$ peanut of each type.

2 The biscuits should be cut into pieces, approximately 1 × 1 cm. Each pupil should be given one piece of each biscuit.

3 **Ventilators/extractor fans should be turned on while food samples are being heated.**

4 **Pupils should wear eye protection.**

Specimen answers

1

Tube	Temperature (°C)		
	Initial	Final	Rise
A	15	38	23
B	15	51	36
C	15	30	15
D	15	48	33

2 A = 23 × 10 = 230 calories; B = 36 × 10 = 360 calories; C = 15 × 10 = 150 calories; D = 33 × 10 = 330 calories (4)

3 A = 23 × 10 × 4.2 = 966 joules; B = 36 × 10 × 4.2 = 1512 joules; C = 15 × 10 × 4.2 = 630 joules; D = 33 × 10 × 4.2 = 1386 joules (4)

Teacher's notes 29

4 (a) There is more fat in the oil-roasted than dry-roasted peanut. Fat has a higher calorific value than carbohydrates or proteins. (2)
(b) There is more fat in the digestive biscuit than in the cream cracker. Fat has a higher calorific value than carbohydrates or proteins. (2)

5 1. The nuts and biscuits were of different sizes and masses. (*Food particles of the same mass, e.g. 0.5 g, should be used.*)
2. Heat is lost in transferring the burning food from the Bunsen burner to its position beneath the tube. (*The burner should be positioned as close as possible to the tube.*)
3. Only some of the heat generated by the burning food goes into the tube to heat the water; the remainder is lost to the air. (*Fit a metal or tinfoil hood to the base of the tube. Burn the food inside this hood.*)
4. Water was heated more strongly at the bottom of the tube than at the top; the temperature at the bottom of the tube was recorded by the thermometer. (*Stir the water; clamp the thermometer so that the bulb is half-way down the tube.*) (3)

18 'E' is for artificial colouring

Investigation: *45–60 mins*

Investigation

MATERIALS

- blue food colouring (labelled A)
- pink food colouring (labelled B)
- blackcurrant squash (labelled C)
- 10–15 cm^3 chromatographic solvent
- 250 cm^3 glass beaker
- 3 strips of silica-coated plastic, each 6.5 × 1.0 cm
- 3 paintbrushes
- ruler, graduated in mm
- pencil

Instructions and notes

1 Blue (brilliant blue + carmoisine) and pink (erythrosine) colourings are available from Sainsbury's, Safeway and other supermarkets. Use a blackcurrant squash that contains carmoisine (E122). Note that Ribena does not contain this compound.

2 As an economy measure, it is possible for 8–12 pupils to share the same bottle of food colouring and paintbrush.

3 The chromatographic solvent contains 9 parts ethanol : 1 part water.

30   Teacher's notes

4 Merck pre-coated TLC-plastic sheets, 20×20 cm, with a layer thickness of 0.2 mm, are suitable for this investigation. Use a guillotine or razor blade to cut the sheets into 6.5×1.0 cm strips. Only if the edge of the strips are straight will the pigments ascend as horizontal bands.

Specimen answers

---

1 (a) blue, $R_f = 0.76$; violet, $R_f = 0.89$ (2)
  (b) red, $R_f = 0.81$ (1)
  (c) violet, $R_f = 0.89$ (1)

2 (a) violet   **(b)** blue   (2)

3 The violet pigment in blackcurrant squash is the same as the violet pigment in solution A. Neither the blue pigment in solution A nor the red pigment in solution B were used to colour the blackcurrant squash.   (2)

4 Removing the silica powder from this part of the strip would prevent upward movement of the solvent.   (1)

5 Once the solvent front has dried it is no longer visible.   (1)

---

19 Natural colouring

Investigation: *40–60 mins*

Investigation

MATERIALS

- leaf of stinging nettle
- culture of *Spirulina* (or another alga)
- 10 cm^3 acetone in a stoppered container
- 2 cm^3 chromatography solvent in a stoppered container
- 2 TLC plastic/silica powder strips[1], each 2×8 cm
- 250 cm^3 beaker
- pestle and mortar
- aluminium foil (to cover beaker)
- paintbrush
- forceps
- pencil (lead)
- coloured pencils (orange, red-pink, yellow, yellow-green, blue-green)
- ruler, graduated in mm
- tissue paper

Instructions and notes

1 The chromatography solvent is a mixture of 55 parts cyclohexane : 45 parts ethyl acetate. **This mixture is highly inflammable and should not be brought close to a naked flame**.

2 If *Spirulina* is not available, any green, blue-green, brown or red alga may be used. Similarly, the leaves of any flowering plant may be substituted for stinging nettle.

Teacher's notes   31

3 The use of pre-coated silica gel sheets is recommended, as separation of pigments is usually more rapid, and consistent R_f values for each pigment are usually obtained among a group of pupils.

Specimen answers

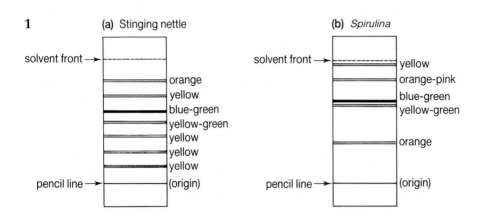

(8)

2 (a) (i) four, (ii) two **(b)** (i) two, (ii) two
(c) blue   (10)

3 Both plants contain two green pigments, one blue-green (chlorophyll a) and the other yellow-green (chlorophyll b).
R_f values for green pigments in stinging nettle:
  yellow-green = 0.70
  blue-green = 0.75

R_f values for green pigments in *Spirulina*:
  yellow-green = 0.84
  blue-green = 0.86
These results suggest some difference between the pigments in the two plants, although they are both chlorophylls and similar in colour. (5

4 1. chemical nature of the absorbent
   2. pore size of the absorbent
   3. air temperature   (2

20 Watching chloroplasts at work

Investigation: 30–50 mins,
March–September

Investigation

MATERIALS

- 50 stinging nettle leaves, freshly picked
- 50 cm³ 10%$^{w/v}$ sucrose solution
- 10 cm³ 3%$^{w/v}$ sodium alginate solution[1] in a stoppered test tube
- 100 cm³ beaker containing 50 cm³ 3%$^{w/v}$ calcium chloride solution
- 100 cm³ beaker
- 5 cm³ plastic syringe

- 100 cm³ measuring cylinder
- petri dish
- glass rod
- spoon
- ruler, graduated in cm
- bench lamp fitted with 100 W bulb
- kitchen mixer
- stop-clock, or watch with a second hand

Instructions and notes

1 The stinging nettle leaves should be freshly picked. Choose young, green leaves. Grind them at high speed for 1–2 minutes in a 10%$^{w/v}$ sucrose solution, to prevent the chloroplasts from bursting.

2 Sodium alginate powder can be mixed with cold water, but may require gentle heating over a Bunsen burner before it will dissolve to produce a homogenous solution.

Specimen answers

1

Distance from lamp (cm)	Time taken for 50% of beads to rise (min)
10	1.5
15	3.0
20	5.5
25	7.0
30	9.0
35	10.0

(6)

2

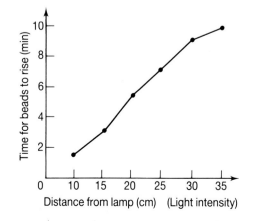

(4)

3 (a) (i)

Reciprocal of distance from lamp $\left(\frac{1000}{d^2}\right)$	Reciprocal of time $\left(\frac{100}{t}\right)$
10	66
4.4	33
2.5	18
1.5	14
1.1	11
0.8	10

(6)

(ii)

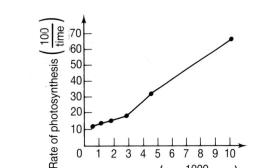

(4)

(b) Figures for the second table were calculated from reciprocals of distance, $\frac{1000}{\text{distance (cm)}^2}$, and time $\frac{100}{\text{time (min)}}$ (3)

4 1. The beads were of different sizes, each with a different number of chloroplasts.
2. Illumination of the beads at the bottom of the measuring cylinder was not uniform. This caused some of the beads to rise before others. (2)

21 Bread-making

Investigation: Part 1 – 70–90 mins
Part 2 – 30–40 mins

Investigation

MATERIALS

Part 1

- 250 g wholemeal flour
- 8 g soya flour
- 8 g activated yeast
- 2 g malt extract (or molasses)
- 2 g sodium chloride
- 150 cm³ warm water
- 5 cm³ sunflower oil
- mixing bowl, wooden spoon
- 250 cm³ beaker
- teaspoon
- Bunsen burner, tripod and gauze

- thermometer
- Clingfilm
- waterbath maintained at 35–40°C
- eye protection

Part 2

- bread dough in mixing bowl
- baking tray
- oven maintained at 220–250°C

Instructions and notes

1 The edible materials are listed in quantities sufficient for making two small loaves or rolls.

2 If individual participation is planned each pupil could be asked to provide a mixing bowl, such as a plastic pudding bowl.

3 If baking trays are not available, bread can be baked in large pie dishes.

4 Dried yeast can be activated by mixing it with sucrose and water in the following proportions: dried yeast: 4 parts; sucrose: 1 part; water: 3 parts. Allow the mixture to stand in a warm place for 6–12 hours.

Specimen answers

1 yeast (*Saccharomyces cerevisiae*) (1)

2 wheat flour; water or milk; fat; sugar; salt. The flour is the source of enzymes. (3)

3 'Leavened' bread has been fermented by yeast, causing the dough to rise. 'Unleavened' bread is not made with yeast, and so it does not rise. (2)

4 'Leavening' promotes CO_2 production, flavour development and texture changes in dough. (3)

5 Alcohol formed during the fermentation of dough is driven off during baking. (1)

22 Wine-making

Investigation: *Part 1 – 30–40 mins*
Part 2 – 60–90 mins
Part 3 – 30–40 mins
Part 4 – 20–30 mins
March–May

Investigation

MATERIALS

Part 1

- 650 g dandelion heads
- 1%$^{w/v}$ sodium hypochlorite solution (for sterilising utensils)

Teacher's notes   35

- large saucepan (3–5 litre capacity)
- plastic bucket (covered)
- access to cooker

Part 2

- soaked dandelion heads
- 1 kg sucrose
- rinds of 2 lemons and 1 orange
- activated wine yeast
- thermometer
- wooden spoon
- large saucepan (3–5 litre capacity)
- plastic bucket (covered)
- access to cooker

Part 3

- 2 g ammonium sulphate
- 2 g ammonium phosphate
- 1%$^{w/v}$ sodium hypochlorite solution (for sterilising utensils)
- gallon jar
- fermentation lock
- muslin bag
- jug

Part 4

- raw wine (in gallon jar)
- 5 g pectinase powder
- 1%$^{w/v}$ sodium hypochlorite solution (for sterilising utensils)
- 4 wine bottles
- 4 screw-caps (or corks)
- filter funnel
- jug
- kitchen towel

Instructions and notes

1 This investigation could take the form of a teacher demonstration.

2 Use dried wine yeast (*Saccharomyces cerevisiae* var *ellipsoideus*), not baking yeast.

3 Dried yeast may be activated by the following procedure. Boil 250 cm^3 water and pour it into a sterilised, washed milk

bottle or similar container. Add the yeast, together with 0.5 g ammonium sulphate and 0.5 g ammonium phosphate. Plug the neck of the bottle with cotton wool. Allow the bottle to stand at 20–25°C for 2–3 days.

4 A 1%$^{w/v}$ solution of domestic bleach may be used if sodium hypochlorite is not available.

5 Pupils should not be allowed to taste any wine prepared in the laboratory.

Specimen answers

1 (a) yeast (*Saccharomyces cerevisiae*) (1)
(b) Yes, but beer is made with a different strain/variety of yeast (*S. cerevisiae*). (2)

2 Sulphur dioxide (SO_2) is a powerful antiseptic. It kills off any bacteria or wild yeasts in the grape juice. If these organisms were not destroyed, the wine might develop an undesirable flavour. (2)

3 Alcohol has a toxic effect on yeast cells. At concentrations in excess of 16 per cent the alcohol inhibits enzymes catalysing the fermentation of sugars. (2)

4 (a) A dry wine is one in which all (or most) of the sugar has been changed into alcohol. This wine is not sweet-tasting. (1)
(b) A fortified wine (e.g. sherry, port, vermouth) is one to which additional alcohol is added after fermentation. This raises the alcohol level to about 20 per cent. (1)
(c) Spirits are wines (e.g. whisky, vodka, rum) that have been distilled. Distillation increases their alcohol level to between 40 and 50 per cent. (1)

23 Malting barley

Preparation: *20–30 mins*
Investigation: *20–30 mins, over 6–8 days*

Preparation

MATERIALS

- 20 g barley grains
- starch/agar/iodine plate
- Clinitest tablets
- Clinitest colour chart
- plastic lunch box
- 100 cm³ beaker, graduated
- flat-bottomed tube, approximately 10 × 2 cm
- 1 cm³ plastic syringe
- pestle and mortar
- scalpel
- forceps
- cotton wool
- sand

Teacher's notes 37

Investigation

MATERIALS

- plastic lunch box with germinating barley grains
- starch/agar/iodine plate with half-grains of barley
- Clinitest tablets
- Clinitest colour chart
- 100 cm³ beaker, graduated
- flat-bottomed tube, approximately 10 × 2 cm
- 1 cm³ plastic syringe
- pestle and mortar
- forceps
- ruler, graduated in mm
- sand

Instructions and notes

1 Instructions for preparing a starch-agar plate are given on page 7. Preparation of a starch/agar/iodine plate involves flooding the surface of a starch-agar plate with iodine solution. The iodine solution is left for 15–20 minutes, then poured away. No further treatment is necessary.

2 Pupils may have difficulty crushing barley grains, especially when they are hard and dry. Mixing the grains with coarse sand makes this task easier.

3 Pupils may have initial difficulties in distinguishing between the embryo- and endosperm-halves of barley grains. Some guidance may be necessary.

Specimen answers

1

Day	Approx. concentration reducing sugar (%)	Mean diameter of starch-free zone (cm)		Area of starch-free zone (cm²)	
		Endosperm half	Embryo half	Endosperm half	Embryo half
1	0.0	0.0	0.0	0.0	0.0
2	0.0	0.4	0.6	0.125	0.28
3	0.0	1.2	1.7	1.13	2.26
4	trace	1.3	2.3	1.32	4.15
5	0.25	1.4	2.8	1.53	6.15
6	0.5	1.5	2.9	1.7	6.60
7	0.75	1.5	3.0	1.7	7.0
8	1.00	1.5	3.1	1.7	7.15
9	0.75	1.5	3.1	1.7	7.15
10	0.5	1.5	3.1	1.7	7.15

(6)

2 (a)

	Daily increase in area of starch-free zone (cm²)	
Day	Endosperm half	Embryo half
1	0.0	0.0
2	0.125	0.28
3	1.175	1.98
4	0.19	1.89
5	0.21	2.00
6	0.17	0.45
7	0.0	0.15
8	0.0	0.0
9	0.0	0.0
10	0.0	0.0

(4)

(b) More amylase is produced by the embryo half-grain, with output reaching a peak after 3–5 days. (2)

3

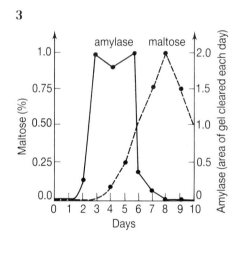

(8)

4 (a) Amylase production peaks after 3–5 days, while maltose production peaks later, after 7–9 days. As the enzyme is a catalyst for the conversion of starch to maltose, enzyme production must occur before malting can take place. (4)

(b)

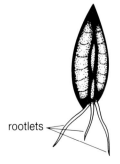

rootlets

(4)

5 1. It was difficult to crush the barley grains effectively, so some maltose may not have been extracted. This means the true maltose concentrations were probably higher than those recorded.
2. The rate at which amylase diffuses into agar is affected by concentration and temperature. This means that any variations in these two factors would affect the rate of diffusion. (2)

Teacher's notes 39

24 Alcohol production

Preparation: *20–30 mins*
Investigation: *30–40 mins*

Preparation

MATERIALS

- 10 g dried yeast
- 100 cm³ 3%$^{w/v}$ sodium alginate solution in a beaker
- 100 cm³ 3%$^{w/v}$ calcium chloride solution in a beaker
- 1dm³ plastic lemonade bottle (or similar fizzy-drink bottle)
- 10 cm³ plastic syringe
- glass rod
- scissors
- silver sand
- cotton wool

Investigation

MATERIALS

- sodium alginate/yeast bead sand column
- 200 cm³ 5%$^{w/v}$ glucose solution in a freezer bag
- 100 cm³ beaker
- 100 cm³ measuring cylinder
- wine hydrometer
- 2 retort stands and bosses
- ring support
- clamp
- pin

Instructions and notes

1 Each pupil should be asked to bring a 1 dm³ lemonade bottle, or similar fizzy-drink bottle.

2 Wine hydrometers are available from stores that stock materials for home-brewing and wine-making. A single hydrometer will serve the requirements of a class.

3 Silver sand is available from garden centres and pet stores. It may need to be washed before it can be used in sand columns. Alternatively, polystyrene beads can be used as packing material.
4 If ring supports are not available, large clamps may be used to hold the plastic bottle in an inverted position.

40 Teacher's notes

Specimen answers

1 (a) $\dfrac{5}{180} \times 92 = 2.55$ g (2)

(b) $\dfrac{5}{180} \times 4480 = 124.4$ cm^3 (2)

2 (a) Some of the glucose is used by the yeast cells to provide energy for growth. This glucose is not converted into ethanol. (2)

(b) If sugar and yeast are kept inside a sealed container, such as a bottle, carbon dioxide will be formed. The volume of gas generated may increase the pressure inside the bottle so much that the glass breaks. An exploding bottle will probably cause damage, and possibly injury. (2)

3 (a) relative density = 1.002 (1)

(b) The product is a glucose solution/ethanol mixture, showing incomplete fermentation of all the glucose to ethanol. (2)

(c) More glucose could be converted into ethanol by passing the product through the apparatus for a second time. (1)

(d) (i) Distillation. (1)

(ii) The boiling point of ethanol is lower than that of water. If an ethanol/water mixture is heated, the distillate becomes enriched with ethanol. This process can be repeated until the distillate consists of pure ethanol. (3)

4 (a) More sodium alginate/yeast beads could be used in the column, which could also be made longer. Both changes should result in more complete conversion of glucose to ethanol. (2)

(b) The same yeast cells can be used over and over again. The yeast cells do not need to be separated from the end product. (2)

25 Making a fermenter

Preparation: *40–90 mins*

Preparation

MATERIALS

- 2 dm^3 plastic bottle, e.g. lemonade bottle
- 40 cm^3 plastic syringe
- hypodermic needle
- glass tubing, 15–20 cm × 35 mm diameter
- PVC translucent tubing, 3–5 mm diameter and 1.0–1.5 cm diameter
- rubber tubing, to fit over glass tubing
- rubber bung, to fit neck of bottle
- rubber bung, to fit probe insertion point
- Bunsen burner
- bradawl (wooden handled)
- Plastic Padding
- clear nail varnish
- paintbrush
- screw clip
- scalpel
- retort stand, boss and ring
- eye protection
- access to fume cupboard

Instructions and notes

1 Heat the plastic bottles only in a fume cupboard. Smoke from the plastic may contain toxic gases.

2 Plastic Padding, available from stores that sell motor accessories, contains hydrogen peroxide in the hardener. Pupils should be warned to avoid skin contact with this product. Not all types of plastic will bond to Plastic Padding, although there is not usually any problem with plastic lemonade bottles. If you do experience problems, an undercoat of aluminium paint will give better adhesion.

3 Coat all joins with clear nail varnish to produce water-tight seals. Check that each join is water-tight by immersion in water.

4 As the plastic lemonade bottles will probably have necks of different diameters, rubber bungs must be chosen that will produce water-tight fits. It is advisable to drill a hole, for the glass tubing, before the corks are distributed. **Furthermore, pupils should be shown how to fit the glass tubing, otherwise hand injuries may occur if the glass breaks during fitting.**

Specimen answers

1 All traces of sodium hypochlorite must be washed away, otherwise micro-organisms, such as bacteria or yeast, may be killed by it.　(1)

2 oxygen/air supply (bubbler); stirrer　(2)

3 1. Stainless steel fermenters are unlikely to burst if subjected to a high internal gas pressure.
2. The steel is unaffected by high temperatures, such as those used for sterilisation, or by acids and alkalis.
3. The smooth, hard surface of a stainless steel vessel is easily cleaned.　(3)

4 Hot water, pumped through the jacket, can be used to sterilise substrates etc. before micro-organisms are introduced. At the end of a batch culture, the fermenter can also be sterilised before the next batch is put into it.　(2)

5 (a) Acetone could be separated from the other solvents by distillation.　(1)
(b) Collect the mixture of gases in a gas tank. Owing to its lower density, hydrogen would rise to the top of the tank, while carbon dioxide would sink to the bottom.　(1)

26　Cell counts

Preparation: *20–30 mins*
Investigation: *30–40 mins, over 6–8 days*

Preparation

MATERIALS

- 2 g dried yeast
- 1 cm³ 5%$^{w/v}$ ammonium phosphate solution
- 50 cm³ 5%$^{w/v}$ sucrose solution in a beaker

42　Teacher's notes

- 2 beakers, each containing 100 cm³ water
- 6 petri dishes
- 5 cm³ plastic syringe
- light meter
- lamp fitted with 100 W bulb
- retort stand, boss and clamp
- glass rod
- dropping pipette
- glass-marking pen
- metre rule

Investigation

MATERIALS

- petri dishes containing yeast suspensions
- light meter
- lamp fitted with 100 W bulb
- retort stand, boss and clamp
- metre rule

Instructions and notes

1 Different types of light meter are available to schools. All types are suitable for use in this investigation, but some modifications in experimental techniques may need to be used with some models.

2 The growth of a yeast population may be extremely slow if cultures are kept at room temperature. Good results can usually be obtained by placing the petri dishes on a radiator or in an incubator at 25–30°C.

Specimen answers

1

Dry mass of yeast (g)	Absorbance (arbitrary units)
0.5	10
1.0	15
1.5	20
2.0	25

(4)

2

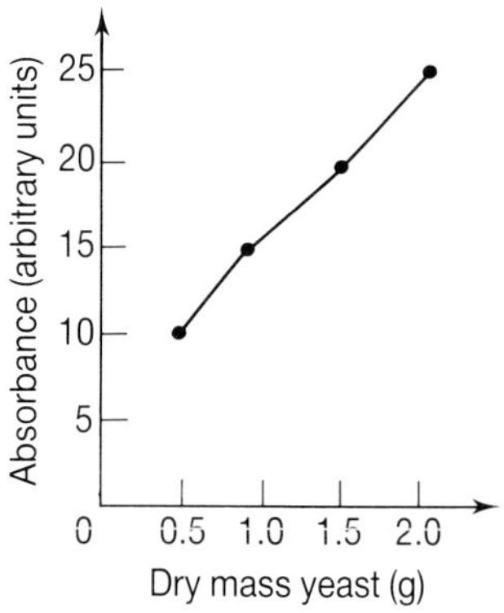

(6)

Teacher's notes 43

| | Absorbance (arbitrary units) ||
Day	+ ammonium phosphate	− ammonium phosphate
0	0	0
2	2	1
4	3	1
6	6	2
8	10	3

(4)

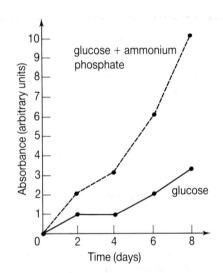

4 Yeast grows and reproduces in both petri dishes, causing a daily increase in light absorbance. However, the growth rate in the dish containing ammonium phosphate is more rapid because this compound supplies the elements nitrogen (N) and phosphorus (P), which are essential nutrients. (6)

5 mass of yeast with ammonium phosphate
= 0.8 g
mass of yeast without ammonium phosphate
= 0.6 g
The absorbance of each culture on day 4 was measured. This was then referred to in the graph drawn for question 2. An increased absorbance of 3 units gives a dry mass of 0.8 g, and of 1 unit a dry mass of 0.6 g. (5)

27 Which antibiotic?

Preparation: 15–20 mins
Investigation: 30–40 mins

Preparation

MATERIALS

- petri dish containing nutrient agar
- Mastring-S[3,4]
- culture of *Bacillus subtilis*[3,4]
- L-shaped glass spreader
- forceps
- glass-marking pen
- clear adhesive tape
- Bunsen burner
- incubator maintained at 25°C

Investigation

MATERIALS

- petri dish containing a Mastring and lawn of *B. subtilis*
- ruler, graduated in mm

Instructions and notes

1 Prepare plates of nutrient agar one day before they are required by pupils. Each pupil will require one plate.

2 Use a freshly-purchased culture of *B. subtilis* as a bacterial source. On the day before pupils prepare their plates, sub-culture the bacterium in a $1.5^{w/v}$ malt extract broth, maintained overnight at 25°C.

3 Ensure that all petri dishes containing bacteria and antibiotic-impregnated discs are sealed with adhesive tape before incubation, and not re-opened.

4 At the end of the investigation, plates containing bacteria and antibiotic-impregnated discs should be destroyed by incineration.

Specimen answers

1

Antibiotic	Diameter of clear zone (cm)	Area of clear zone (cm^2)
chloramphenicol	3.4	9.07
erythromycin	2.5	4.90
novobiocin	2.2	3.79
penicillin G	0.1	0.78
streptomycin	0.0	0.0
tetracycline	1.8	2.54

(6)

2 Chloramphenicol; erythromycin; novobiocin; tetracycline; penicillin G; streptomycin.　　(6)

3 (a) $\dfrac{9.07}{3.79} = 2.39$

(b) $\dfrac{9.07}{2.54} = 3.57$　　(4)

4 Erythromycin, because it is the second most active antibiotic against the bacterium.　　(2)

5 (a) Sealing prevents any antibiotic-resistant bacteria from escaping into the air.　　(1)
(b) If plates are heated with agar at the bottom, moisture collects on the lid and drops of water drip down on the gel, causing the bacteria to spread.　　(1)

Teacher's notes　45

28 Daily urine output

Investigation: 48 hours

Investigation

MATERIALS

- 250 cm³ measuring cylinder

Instructions and notes

1 There are usually some pupils in each class who are prepared to provide data. They should be encouraged. Guidance should be given on sterilising measuring equipment and pupils should take individual responsibility for cleaning the measuring equipment they use.

Specimen answers

1
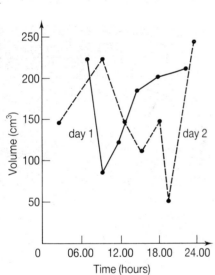
(4)

2 **(a)** 1010 cm³ on day 1; 1040 cm³ on day 2 (2)

(b) $\dfrac{2050}{2} = 1025$ cm³ (1)

(c) $\dfrac{1025}{24 \times 60} = 0.71$ cm³ (1)

3 **(a)** $220 + 80 + 120 + 180 + 200 + 220 + 150 + 110 + 140 = 1420$ cm³ (1)

(b) $210 + 140 + 230 + 50 = 630$ cm³ (1)

4 More urine is produced between 06.00 and 18.00 than between 18.00 and 06.00 h. (1)

5 Take a sample of 100 or more adults, 50% male:50% female. Hourly intake of 200 cm³ water. No food. Hourly urination. Measure and record urine output. Continue with the investigation over a 24 h period. (4)

29 Drinking and driving

Investigation: 5–10 mins

Investigation

MATERIALS

- whisky glass
- wine glass
- beer glass
- 100 cm³ measuring cylinder

Instructions and notes

Pupils could be asked to measure and record the volume of glasses in their own homes. Data could then be analysed during a class period. Alternatively, one set of drinking glasses could be used per class.

Specimen answers

1 The whisky glass held 75 cm³, wine glass 175 cm³ and beer glass 250 cm³. (1)

whisky glass: $\frac{75}{100} \times 35 = 26.25$ cm³ alcohol (1)

wine glass: $\frac{175}{100} \times 12 = 21.00$ cm³ alcohol (1)

beer glass: $\frac{250}{100} \times 4 = 10$ cm³ alcohol (1)

2 The volume of alcohol in the man's blood must not exceed $5\,000 \times 0.01 = 50$ cm³ alcohol. (1)

whisky: $\frac{50}{26.25} = 1.90$ (1 glass) (1)

wine: $\frac{50}{21.00} = 2.38$ (2 glasses) (1)

beer: $\frac{50}{10} = 5.0$ (5 glasses) (1)

3 It is assumed that all of the alcohol consumed is absorbed into the blood stream. (1)

4 In general, women cannot drink as much alcohol as men before they exceed the legal limit. (1)

30 Variation in a human population

Investigation: 60–90 mins

Investigation

MATERIALS

- vertical scale, graduated in cm, for measuring height
- metric scales, for measuring mass

Instructions and notes

1 The simplest device for measuring height is a metre rule or tape measure fixed in a vertical position to a wall with clear adhesive tape.

2 If scales calibrated in metric units are not available, those calibrated in stones and pounds may be used. Divide the number of pounds by 2.2 to convert to kilograms (kg).

Specimen answers

1 (a)

	No. pupils showing each characteristic		
Character	Dominant	Recessive	Total
eye colour	8	5	13
earlobe attachment	12	1	13
tongue-rolling	9	4	13
freckling	9	4	13
thumb dominance	8	5	13

(5)

(b) For each characteristic listed, the dominant phenotype was more numerous than the recessive. (2)

(c)

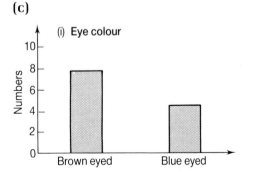

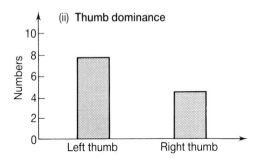

(4)

48   Teacher's notes

(d) (i) $\frac{5}{13} \times 100 = 38\%$

(ii) $\frac{8}{13} \times 100 = 61.5\%$ (2)

(e)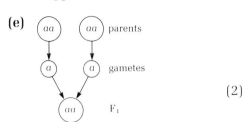
(2)

(f)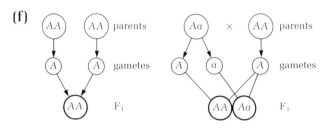

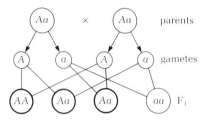

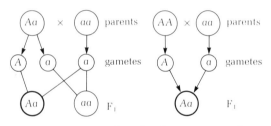

(Dominant F_1 individuals are ringed in heavy type.) (5)

2 (a) 154, 158, 158, 160, 161, 164, 164, 166, 167, 170, 171, 172, 196 cm (2)

(b) $154 + 158 + 158 + 160 + 161 + 164 + 164 + 166 + 167 + 170 + 171 + 172 + 196 \div 13 = 166$ cm (2)

(c)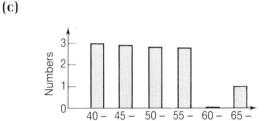
(6)

3 (a) 41, 41, 41.5, 45, 45, 48, 50, 52, 53, 55.5, 56, 57, 69 kg (2)

(b) $41 + 41 + 41.5 + 45 + 45 + 48 + 50 + 52 + 53 + 55.5 + 56 + 57 + 69 \div 13 = 50$ kg (2)

(c)

(6)

31 Fingerprint analysis

Investigation: 40–50 mins

Investigation

MATERIALS

- 5 girls and 5 boys (subjects)
- ink pad
- fingerprint ink[3,4]
- pen or pencil
- white paper

Teacher's notes 49

Instructions and notes

1 Fingerprint ink, readily removed by washing with water, is available from most biological suppliers.

2 Fingerprints, made on a clear sheet of plastic, can be projected onto a screen from an overhead projector.

3 In certain instances pupils may experience difficulties in identifying pattern types. This is particularly likely to occur if the subjects are young children, middle school pupils with small hands, or those who have injured the tips of their fingers. The provision of a hand magnifying glass, with a magnification of × 5 or × 10, may be necessary.

Specimen answers

1 (a) (i) *Left hand*

Digit:	1	2	3	4	5	
Print pattern:	W	W	W	W	UL	(2)

(ii) *Right hand*

Digit:	1	2	3	4	5	
Print pattern:	W	W	W	UL	W	(2)

(b) none (1)

2

Fingerprint pattern	No. of each type		Percentage in whole population
	Girls	**Boys**	
plain arch	2	1	3
tented arch	3	2	5
ulnar loop	34	31	65
radial loop	4	3	7
whorl	7	13	20

(5)

3 (a) (i) ulnar loops, (ii) ulnar loops (2)
   (b) (i) plain arches, (ii) plain arches (2)

4 (a) (i) $\frac{34}{50} \times 100 = 68\%$; (ii) $\frac{31}{50} \times 100 = 62\%$;

     (iii) $\frac{65}{100} \times 100 = 65\%$ (3)

   (b) $\frac{13}{50} \times 100 = 26\%$ (1)

5 arches and loops (2)

6 Fingerprint analysis could be used as a diagnostic tool in medicine. For example, a high frequency of certain pattern types may indicate genetically transmitted disorders, a predisposition towards certain diseases, or abnormalities in chromosome number. (2)

7 The sample size of five girls and five boys was too small to produce accurate results. The sample should be increased to 100 girls and 100 boys. (3)

50   Teacher's notes

32 How exercise affects your heart

Investigation: *30–50 mins*

Investigation

MATERIALS

- gymnasium bench or stool, 25–35 cm in height
- metronome
- stop-clock, or watch with a second hand

Instructions and notes

1 Any pupils known to suffer from cardiovascular or respiratory problems should be excluded from this investigation.

2 All members of the class should exercise simultaneously in time with the metronome. Individual records of heart rates should be kept, and the relevant results may then be pooled by making individual entries in a table written on the blackboard.

3 You may need to demonstrate methods for detecting and recording pulse rates.

Specimen answers

1

Time (min)		Rate of heart beat (beats/min)
−2	preparation	70
−1		72
0		not measured
1	exercise	not measured
2		not measured
3		not measured
4		126
5		97
6		88
7	recovery	76
8		73
9		71
10		71

(4)

Teacher's notes   51

2

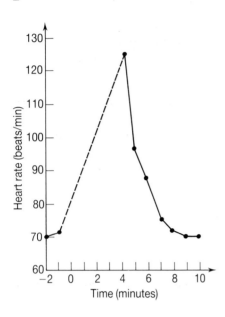

(6)

3 (a) There was a small rise in heart rate before the exercise began. (1)
(b) This increase is caused by adrenalin, released into the bloodstream in response to increased stress through anticipation etc. (1)

4 (a) 189   **(b)** 115   (2)

5 (a) 6 minutes   **(b)** 14 minutes   (2)

6 The higher the pulse rate during exercise and the longer the recovery period, the less efficient the circulatory system. This means that a pupil with a heart rate of 115 after exercise and a 6 minute recovery period has a more efficient circulatory system than one with a heart rate of 189 and a recovery period of 14 minutes. (2)

7 Yes. The 'step test' indicates the relative efficiency of the circulatory system in pupils of the same age. It therefore identifies those most at risk from diseases of the cardiovascular system (high blood pressure, heart attack, stroke) in later life. (2)

33 Heart and breathing rates during exercise

Investigation: *40–60 mins*

Investigation

MATERIALS

- gymnasium bench
- metronome
- stop-clock, or watch with a second hand
- graph paper

52   Teacher's notes

Instructions and notes

1 Pupils with circulatory or respiratory problems should not take part in this investigation.

2 If a gymnasium bench is not available, alternative forms of exercise, such as running, sit-ups, squat-jumps, etc. could be used.

3 An exercise bicycle, operated under supervision, could also be used to obtain results. Some exercise bicycles have scales showing resistance and miles per hour. If you have a bicycle of this type ask pupils to pedal at a rate of 5−10 miles per hour, and adjust the resistance so that pupils obtain meaningful results. If the bicycle has neither of these features, ask pupils to pedal at 30−40 revolutions per minute.

Specimen answers

1 (a)

Duration of exercise (minutes)	1	2	3	4	5	6	7	8	9	10
Breaths per minute	17	19	21	24	27	28	33	36	42	44

(4)

(b)

Duration of exercise (minutes)	1	2	3	4	5	6	7	8	9	10
Heart beats per minute	57	65	70	73	79	82	91	101	114	119

(4)

2 (a)

(5)

(b) The graph is a straight line (linear graph). (1)

(c) There is a direct relationship between ventilation and heart rate, suggesting that the two processes are linked, or correlated (as indeed they are by the autonomic nervous system). (2)

3 1. Each person could produce two or more sets of results for heart rate and ventilation rate. Mean values could then be calculated and plotted. (2)

2. Two or more individuals could be asked to produce sets of results. Again, mean values could be calculated and plotted. (2)

Teacher's notes **53**

34 Lung capacity

Investigation: *40–60 mins*

Investigation

MATERIALS

- 10 subjects
- hand (pocket) spirometer[4]
- 10 individual mouthpieces
- metric scales, for measuring body mass

Instructions and notes

1 Individual mouthpieces are supplied with the spirometer. These can be collected in a 1 dm³ beaker containing a 1% solution of domestic bleach at the end of the investigation.

2 An alternative to the hand spirometer for measuring vital capacity is a large bell jar which has been calibrated and a vacuum pump.

3 After it has been used by several pupils, saliva may collect in the central cavity of the spirometer. This should be removed with tissue paper and the walls of the cavity wiped with a disinfectant (e.g. diluted domestic bleach).

Specimen answers

1 (a) (This data is for boys aged 14–15 years.)

Subject no. and name	Vital capacity (cm³)	Body mass (kg)	Vital capacity (cm³)/ body mass (kg)
1 E. Brown	3250	57	57.0
2 S. Ling	4100	55.5	73.8
3 J. Smith	3100	45	68.8
4 M. Richards	2800	41.5	67.5
5 P. Robinson	3800	48	79.2
6 F. de Souza	4000	69	57.9
7 K. Murray	3800	45	84.4
8 L. Roberts	4100	56	73.2
9 D. Grey	3200	41	78.0
10 R. Price	3100	41	75.6

(6)

(b) Data included in the table, after calculating vital capacity/body mass ratio. (5)

(c) (i) The value obtained from this equation is equal to approximately 1.0. Any figure below 1.0 would indicate that a subject might rapidly become short of oxygen during vigorous exercise. Conversely, figures above 1.0 would indicate better than average oxygen supply. (2)

(ii) As the lungs supply oxygen to the tissues, the ratio lung volume:body mass indicates relative amounts of oxygen available to each unit mass (e.g. kg) of tissue. The higher the ratio, the more oxygen is available to the tissues. (2)

2 (a)

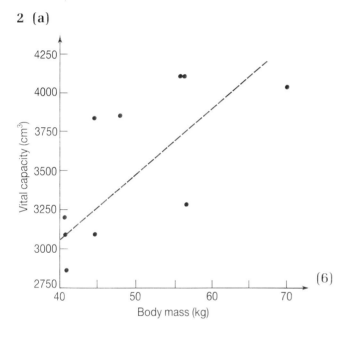

(6)

(b) There was probably a general tendency for vital capacity to increase directly with increasing body mass. Even so, there was considerable variation among individuals with similar body masses. For example, two subjects with identical body masses (45 kg) had vital capacities of 3100 cm^3 and 3800 cm^3 respectively. (4)

35 Measuring grip strength

Investigation: *10–15 mins*

Investigation

MATERIALS

- dynamometer[3,4]

Instructions and notes

When measuring grip strengths it is important that all pupils should hold the dynamometer in the same way, with the arm kept straight at the elbow.

Specimen answers

1

Hand	Grip strengths (kg)	Mean
right	29, 36, 31, 41, 29, 34, 36, 38, 31, 27, 37, 29, 29, 33, 30, 33, 37, 28	32.6
left	27, 33, 30, 38, 27, 31, 32, 38, 33, 24, 34, 27, 21, 31, 29, 31, 35, 27	30.7

(6)

2 (a) *Right-handed people:*

Hand	Grip strengths (kg)	Mean
right	31, 36, 31, 27	31.25
left	30, 32, 30, 24	29.0

Left-handed people:

Hand	Grip strengths (kg)	Mean
right	38, 32, 30, 31	32.75
left	38, 36, 33, 33	35.00

(4)

(b) In right-handed people, grip strength of the right hand was stronger than in the left. Conversely, in left-handed people, grip strength of the left hand was stronger than that of the right. Results are consistent with the view that the strength of muscles is related to the frequency of their use. (4)

3 diabetes; muscular dystrophy; cystic fibrosis. (2)

4 (a) (i) $\frac{1}{24} \times 100 = 4.16\%$ stronger at 11

(ii) $\frac{5}{50} \times 100 = 10\%$ stronger at 17 (2)

(b) (i) $\frac{30}{25} \times 100 = 120\%$ stronger in right hand

(ii) $\frac{26}{24} \times 100 = 108.3\%$ stronger in left hand (2)

36 Obesity

Investigation: *40–60 mins*

Investigation

MATERIALS

- 10 subjects
- skin-fold callipers[4]
- vertical scale, graduated in cm, for measuring height
- metric scales, for measuring body mass

Instructions and notes   Several different types of skin-fold callipers are available from manufacturers. In order to obtain consistent results, it may be necessary to demonstrate how the callipers should be used.

Specimen answers

1 (a)

Subject no. and name	Thickness of triceps skin-fold (mm)	Height (cm)	Body mass (kg)	Height (cm)/ body mass (kg)
1 K. James	7	172	57	3.01
2 L. Johnson	9	170	55.5	3.06
3 F. Williams	7	161	45	3.57
4 V. Pearce	4	158	41.5	3.80
5 K. Jenkins	8	164	48	3.41
6 L. Simmons	5	196	69	2.84
7 A. Singh	7	158	45	3.51
8 P. O'Meara	9	164	52	3.15
9 D. Frost	5	160	41	3.90
10 K. Bartlett	8	171	56	3.05

(5)

(b) The fifth column of the table has been calculated from the formula
height (cm) ÷ body mass (kg).   (5)

(c) It would be more meaningful to divide body mass by height. This means a better formula would be:

$$\frac{\text{body mass (kg)} \times 3}{\text{height (cm)}}$$

The formula would produce a figure of approximately 1.0 for those of normal build, higher than 1.0 for those who are tall and thin, and lower than 1.0 for those with a tendency towards obesity.   (2)

2 (a)

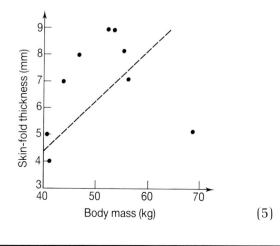

(5)

(b) There is probably a general tendency for the thickness of the triceps skin-fold to increase with body mass. Even so, there is considerable individual variation, with individuals of almost identical body mass showing differences in the thickness of their skin-folds.   (4)

3 (a) One effect of the sex hormone testosterone in boys is to cause the redistribution of subcutaneous fat during adolescence. The initial effect of this hormone is to cause a loss of subcutaneous fat, which builds up again during late adolescence. (3)

(b) In girls a different sex hormone, oestrogen, regulates redistribution of subcutaneous fat during adolescence. (2)

4 An obese person should be encouraged to eat a balanced diet, but to reduce their total calorie intake, especially their intake of carbohydrates and fats. This should make the body use up its fat stores, leading to a loss in mass. In addition, an obese person should take more exercise in an attempt to burn up more calories, and consult a doctor if any of these measures fail to cause a loss in weight. (4)

37 Successful seed germination

Preparation: *30−40 mins*
Investigation: *30−40 mins*

Preparation

MATERIALS

- seeds of 5 flowering plants (e.g. stock, pansy, polyanthus, arctic poppy, French marigold)
- 200 cm^3 1%$^{w/v}$ Benlate solution
- garden soil
- seed compost
- chick grit
- 10 15−25 cm flower pots
- 10 large plastic bags, e.g. 30 × 40 cm
- spoon
- string
- glass-marking pen
- garden trowel

Investigation

MATERIALS

- 5 flower pots containing soil-grown seedlings
- 5 flower pots containing compost-grown seedlings

58 Teacher's notes

Instructions and notes

1 This investigation should be carried out using seeds of five different species of flowering plant. Pupils will require 100 seeds of each species. It is best to avoid very small seeds, such as those of salvias.

2 John Innes or Levington seed composts provide a suitable sowing medium for the seeds of most species. The seeds of acid-loving (lime-hating) plants, such as heathers and rhododendrons, however, should be sown in ericaceous seed compost, or a mixture of peat and sand.

Specimen answers

1

Name of seed	No. of seedlings/pot	
	Garden soil	Seed compost
stock	38	46
pansy	33	47
polyanthus	46	48
arctic poppy	14	48
French marigold	41	47

(5)

2 (a) Percentage germination in garden soil: stock (76); pansy (66); polyanthus (92); arctic poppy (28); French marigold (82).
(b) Percentage germination in seed compost: stock (92); pansy (94); polyanthus (96); arctic poppy (96); French marigold (92). (5)

3 The general effect of sowing seeds in compost and treating them with fungicide was to increase the percentage germination. Even so, the effect was more marked in some species than in others, ranging from a 4% increase in polyanthus to a 68% increase in arctic poppy. (3)

4 (a) Germinating the seeds in a plastic bag retained moisture, kept out animals and protected the young plants from wind, extremes of temperature, etc.

(b) The main disadvantage is that fungi, bacteria, algae, mosses, etc. are more likely to grow at the soil surface in the more moist atmosphere enclosed by the bag.
(c) The growth of fungi could be prevented by using a fungicide. An algicide would prevent the growth of algae. (5)

5 1. Duplicate the number of pots/seeds for each treatment.
2. Treat both soil-sown and compost-sown seeds with (i) water and (ii) Benlate solution. (It isn't clear from the design of the investigation if the compost or fungicide has most effect on the percentage germination.) (2)

38 New bulbs from old

Investigation: *Part 1 – 30–40 mins*
Part 2 – 10–15 mins

Investigation

MATERIALS

Part 1

- daffodil bulb
- 200 cm³ 50%$^{w/v}$ ethanol solution in a 250 cm³ beaker
- Benlate (a fungicide)
- plastic bag, half-filled with moist vermiculite
- 250 cm³ beaker
- white tile
- forceps
- scissors
- string
- scalpel
- glass rod
- Bunsen burner
- plastic gloves
- eye protection
- access to fume cupboard and greenhouse

Part 2

- plastic bag containing young bulbs
- 2 flower pots
- sterile potting compost, e.g. Levington compost

Instructions and notes

1 This is a two-part activity. The second stage is carried out 10–12 weeks after the first.

2 A 1%$^{w/v}$ solution of domestic bleach preferably without added detergent may be used in place of ethanol solution to sterilise the cut surfaces of bulbs. This treatment, however, may not produce such satisfactory results.

3 Benlate is a commercial preparation of the fungicide benomyl. It is available from most garden centres, supermarkets, etc.

4 Horticultural vermiculite, available from garden centres and biological suppliers, should be prepared overnight. Put dry vermiculite into a bucket and add approximately one-third of its volume of water. Wring out excess water from the soaked vermiculite before putting 3–4 handfuls into each plastic bag.

60 Teacher's notes

Specimen answers

---

1 (a) daffodil; snowdrop; hyacinth (1)
  (b) onion; leek (1)

2 Ethanol (alcohol) solution is a powerful antiseptic, which kills off any bacteria and fungi present. Without this treatment bacteria and fungi would attack the cut surfaces. (1)

3 $\dfrac{50}{10} \times \dfrac{100}{50} = £10$ per bulb (3)

4 commercial value of
  new bulbs $\quad = 100 \times 5 \times 5 \times 7$
  $\qquad\qquad\qquad = £17\ 500$
  cost of original bulbs $= £500$
  profit $\qquad\qquad = £17\ 500 - 500$
  $\qquad\qquad\qquad = £17\ 000$ (4)

---

39 New plants from cuttings

Investigation: *Part 1 – 20–30 mins*
(Nov–Dec)
Part 2 – 20–30 mins
(Jan–Feb)
Part 3 – 40–50 mins
(March–April)

Investigation

MATERIALS

Part 1

- packet of geranium (pelargonium) seeds
- root tuber of dahlia
- Benlate solution (0.1%$^{w/v}$)
- potting/seed compost
- 2 15–25 cm flower pots
- 2 large plastic bags
- string
- spoon
- garden trowel
- natural light, e.g. window, or artificial light, e.g. lamp

Part 2

- pot containing geranium (pelargonium) seedlings
- pot containing dahlia tuber with buds/shoots
- Benlate solution (0.1%$^{w/v}$)
- potting compost
- 20 15–25 cm flower pots
- 20 plastic bags
- string
- spoon
- garden trowel
- scalpel
- natural light, e.g. window, or artificial light, e.g. lamp

Teacher's notes 61

Part 3

- potted geranium (pelargonium) plants
- potted dahlia plants
- potting compost
- rooting powder
- 50 15–25 cm flower pots
- forceps
- scalpel
- garden trowel

Instructions and notes

1 The aim of this activity, which involves seed germination and vegetative propagation from young plants, is to produce large numbers of mature plants for bedding.

2 Continuous illumination may be used to encourage the formation of lateral shoots. A mounted fluorescent lamp, raised 15–20 cm above the bench by 2–3 bricks at each end, will provide adequate light intensity for growth. Lamp bulbs of 85 or 125 W are the most suitable for this investigation. Alternatively, commercial 'daylight' bulbs are available from some biological suppliers, and garden centres that specialise in greenhouse fittings.

3 The activity could be extended to include *Antirrhinum*, *Impatiens*, *Coleus*, *Petunia* and other frost-tender species.

4 Success in this investigation depends on making an early start in the autumn and maintaining high light intensities and temperatures throughout the winter.

Specimen answers

1 **(a)** carnation; geranium; *Impatiens* (1)
  (b) *Rhododendron*; heather; blackcurrant (1)

2 Most cuttings are taken in the spring or early summer. At this time of year growth is most rapid. New shoots and roots are more likely to form than in other seasons of the year. (2)

3 Auxin is the hormone used in rooting powders. It is applied to the base of the cut stem immediately before planting. (2)

4 Cuttings must develop a root system before they can absorb water from the soil. Reducing the number of leaves per cutting reduces water loss so that the cuttings do not dry out before their root systems have become established. (2)

5 The leaves can be sprayed with an anti-transpirant to reduce water loss. Alternatively, pots containing cuttings can be put inside plastic bags. Fungicides can be used to prevent attacks from pathogenic fungi. (2)

40 Rooting cuttings in a gel

Preparation: *20–30 mins*
Investigation: *5–10 mins, over 30 days*
April–September

Preparation

MATERIALS

- 0.01 g indole acetic acid (IAA)
- 10 g bacteriological agar powder
- 1 cm³ bleach solution
- ethanol
- 1 g Benlate
- 5 Sach's water culture tablets[3,4]
- 1 dm³ pyrex beaker
- 500 cm³ pyrex beaker
- 100 cm³ beaker
- test tube fitted with a rubber bung
- 4 transluscent plastic cups
- aluminium foil
- scissors
- Bunsen burner, tripod and gauze
- pestle and mortar
- glass rod
- glass-marking pen
- eye protection

Investigation

MATERIALS

- 4 cuttings of geranium (pelargonium)
- 4 cuttings of mint
- 2 agar cups (– mineral salts)
- 2 agar cups (+ mineral salts)
- pencil
- scalpel

Instructions and notes

1 Cuttings of the following plants are suitable for rooting in gels: African violet; *Chrysanthemum*; *Coleus*; *Echeveria*; *Fuchsia*; *Gloxinia*; *Peperomia*; *Sansevieria*; spider plant; watercress.

2 Bleach is added to the gel as an antibacterial agent. Use a bleach solution that does not contain soap or detergent. Alternatively, use a 10%$^{w/v}$ solution of sodium hypochlorite.

3 Heat molten agar in pyrex beakers, stirring continuously. Failure to do this may result in local heating, which may cause the beaker to crack.

63

Specimen answers

1

Day	Geranium No. roots	Geranium No. new leaves	Mint No. roots	Mint No. new leaves
0	0	0	0	0
2	0	0	0	0
4	0	0	0	0
6	0	0	0	0
8	0	0	1	0
10	0	0	1	0
12	1	0	2	0
14	1	0	2	0
16	3	0	3	2
18	3	0	4	2
20	4	0	4	2
22	4	0	5	4
24	4	2	5	4
26	5	2	6	4
28	5	2	7	6
30	6	2	9	6

(7)

2 Mineral salts are an essential raw material for photosynthesis. Adding mineral salts to the gel stimulated the growth of both roots and shoots. Those cuttings supplied with mineral salts were larger, with more roots and leaves, than those grown in agar alone. (4)

3 1. The gels, especially those containing mineral salts, were cloudy.
2. After more than 15–20 days the gels began to dry out. (2)

4 An agar gel, containing bleach and Benlate, but no IAA or mineral salts. (2)

41 Micropropagation

Investigation: Part 1 – 50–60 mins
Part 2 – 50–60 mins

MATERIALS

Part 1

- dandelion plant with tap root
- 20 cm³ 1%$^{w/v}$ Benlate solution in a petri dish

64 Teacher's notes

- 100 cm^3 beaker with a 2 cm depth of agar gel
- scalpel
- forceps
- Clingfilm
- light source

Part 2

- dandelion segments with leaves
- 20 cm^3 1%$^{w/v}$ Benlate solution in a petri dish
- 4 100 cm^3 beakers, each with a 2 cm depth of agar gel
- scalpel
- forceps
- Clingfilm
- light source

Instructions and notes

1 It is possible to teach this technique throughout the year, as dandelion roots are always present in the soil.

2 Prepare the agar from a refined powder, such as bacteriological agar. Add 1 g agar powder to 100 cm^3 water, boil, and pour the molten agar to a depth of 2 cm. Allow time for the agar to cool and harden.

3 Root segments may be exposed to artificial light, such as that produced by a mounted fluorescent lamp supported by bricks at a height of 15–20 cm above the bench.

Specimen answers

1 Micropropagation is a rapid method of producing new plants from stem, leaf or root tissue. Large numbers of uniform, high quality plants can be produced in a relatively short time. The small plants produced by this technique are easily stored or transported. Furthermore, micropropagation from apical meristems can eliminate viruses and other pathogens. The main disadvantages of micropropagation is that plant material must be set up and maintained under aseptic conditions. If aseptic conditions are not maintained the plant material is rapidly killed by bacteria and fungi. (3)

2 **(a)** Auxin stimulates cell elongation and promotes root formation in tissue explants. (1)
(b) Gibberellin also stimulates cell elongation. (1)
(c) Cytokinin promotes cell division. (1)
(d) Agar retains moisture, providing a suitable environment for tissue growth. (1)
(e) Sodium hypochlorite is an antiseptic, used to sterilise tissues before they are placed in containers on agar. (1)

3 The term 'clone' refers to identical genetic copies of an individual. Micropropagation produces such copies, or clones, of the parent plant. (2)

42 Pollination

Investigation: *Part 1 – 10–15 mins
(June–July)
Part 2 – 10–15 mins
(Dec–Jan)
Part 3 – 20–30 mins
(March–April)
Part 4 – 10–15 mins
(June–July)*

Investigation

MATERIALS

Part 1

- polyanthus or primrose seeds in 2 different varieties/strains, A and B
- 2 seed boxes
- seed compost
- garden trowel

Part 2

- polyanthus or primrose plants in 2 varieties/strains, A and B
- Benlate solution ($1\%^{w/v}$)
- potting compost
- 10 15–25 cm flower pots
- 10 plastic bags
- string
- spoon
- garden trowel

Part 3

- pin-eyed flowering plant of variety/strain A
- thrum-eyed flowering plant of variety/strain B
- scissors
- plastic bag
- string

Part 4

- pollinated plant bearing seed pods

Instructions and notes

1 Part 1 of this activity involves raising plants from seeds. Clearly, if plants are readily available this stage can be omitted.

2 It is suggested that the varieties/strains chosen should bear flowers differing markedly in colour, e.g. white/red, or yellow/blue.

Specimen answers

1 insects; slugs and snails (2)

2 (a) The plastic bags prevented insects and molluscs from cross-pollinating the flowers. This could have happened both before and after the intended cross-pollination had been carried out. (1)
(b) Petals of the thrum-eyed flowers were removed to expose the stamens, half-way down the corolla tube. (1)

3 (a) The white-flowered allele is dominant to red-flowered. (1)
(b) The red-flowered allele is dominant to white-flowered. (1)
(c) The red- and white-flowered alleles are incompletely dominant. (1)

(d) One parent plant was a heterozygote (Aa) and the other was homozygous recessive (aa). (1)

4

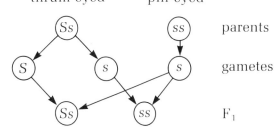

The F_1 generation consists of pin-eyed and thrum-eyed plants in roughly equal numbers. (2)

43 Longer life for cut flowers

Preparation: 15–20 mins
Investigation: 30–40 mins, over 5–9 days

Investigation

MATERIALS

- 50 daffodils or other flowers (freshly picked)
- 5 500 cm³ conical flasks
- 50 g sucrose
- 0.5 g sodium salicylate
- packet of commercial flower preservative, e.g. Chrysal
- hammer
- glass rod
- glass-marking pen

Instructions and notes

1 The sample answers were obtained using daffodil flowers. Other types of flowers may be used, depending on availability. If other species are used, you must provide guidelines on the point at which a flower is considered to be 'dead'.

2 It is suggested that one investigation should be carried out per class. Destructive sampling of flowers should be avoided.

3 Flower preservatives, such as Chrysal, are available from florists and supermarkets.

Teacher's notes 67

Specimen answers

1

Flask no.	No. living flowers/flask									
	Day 0	1	2	3	4	5	6	7	8	9
1	10	10	10	10	2	0	0	0	0	0
2	10	10	10	10	4	2	0	0	0	0
3	10	10	10	10	10	8	4	1	0	0
4	10	10	10	10	2	0	0	0	0	0
5	10	10	10	10	7	3	1	0	0	0

(8)

2

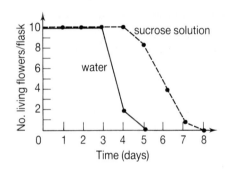

(4)

3 (a) Crushing the stems prolonged the vase-life of some flowers by 1–2 days. (1)
(b) (i) Cut a transverse section of the stem and look at it under a microscope to find out if the xylem vessels are blocked.
(ii) Look at a drop of water under a microscope to find out if it contains bacteria, algae and protozoa. (2)

4 (a) Sucrose (10%$^{w/v}$ solution) was the most effective in extending the vase-life of cut daffodils. (1)
(b) Sodium salicylate had little, if any effect on the vase-life. (1)
(c) Chrysal, a commercial preservative for cut flowers, prolonged the vase-life of some flowers by anything from 1–3 days, but its effects were not as marked as a 10%$^{w/v}$ sucrose solution. (1)

5 Crush the stems and stand the daffodils in a 10%$^{w/v}$ sucrose solution. (2)

44 Dispersing weed seeds

Investigation: 40–60 mins

Investigation

MATERIALS

- ripe fruiting head of dandelion with 10 or more fruits
- metre rule
- scalpel
- magnifying glass
- retort stand, boss and clamp
- stop-clock, or watch with a second hand

Instructions and notes

1 The chief aim of this investigation is for pupils to devise their own experimental techniques.

2 It is not essential to supply pupils with an intact fruiting head of dandelion. An incomplete head, with 10 or more fruits, will suffice. The fruiting head can be supplied in a petri dish.

Specimen answers

1 Use the scalpel to cut the receptacle into 8 equal segments. Count the number of pits in one segment. Multiply the number of pits in one segment by the number of segments (e.g. 22 × 8 = 176). (4)

2 176 × 12 = 2112 (2)

3
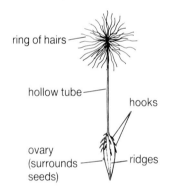
(3)

4 (a) Clamp the ruler, in a vertical position, 1 m above the bench. Experiment with (i) intact fruit, (ii) trimmed parachute + fruit, (iii) fruit only, (iv) parachute only. Drop each, in turn, from 1 m above the bench. Time two or more descents. Record results. (4)

(b)

	Time (sec) for 1 m descent		
Condition of fruit	1st descent	2nd descent	Mean
intact fruit	2.67	2.41	2.59
trimmed parachute	1.55	1.75	1.65
fruit only (no parachute)	0.75	0.87	0.81
parachute only	6.31	8.23	7.27

(4)

5 (a) wind; relative humidity of the air (1)
 (b) (i) The parachute keeps the seed buoyant in dry conditions, delaying its rate of descent.

 (ii) After the fruit has landed, the hooks cling to soil or plants, preventing re-launching. (2)

Teacher's notes 69

45 Trapping and killing slugs

Preparation: *Part 1 – 10–15 mins*
Part 2 – 20–30 mins
Investigation: *40–50 mins, over 3–4 days*
March–July

Preparation

MATERIALS

Part 1

- slug pellets
- 5 small plastic bags
- plastic gloves
- top-pan balance
- glass-marking pen

Part 2

- slug pellets (in plastic bags)
- 5 1 m² squares of carpet (or 10 0.5 m² squares of newspaper)
- 1 m² quadrat
- 5 bricks or large stones (10 if newspaper is used)
- 20 sticks
- hammer
- plastic gloves
- cultivated soil, e.g. garden soil

Instructions and notes

1 Slug pellets are poisonous to humans. If pupils are to weigh and scatter the pellets, they should wear plastic gloves.

2 As an alternative to an area of cultivated garden soil, it may be possible to carry out the investigation on a lawn, or any piece of waste ground covered by short vegetation.

3 Each day, after counts have been made, the pupils should wear plastic gloves to collect and remove all the slugs that have been trapped. Failure to do this may result in inaccurate counts.

Specimen answers

1 (a) *carpet/newspaper traps:*

	No. slugs trapped					
Day	**Trap no. 1**	2	3	4	5	**Total catch**
1	4	5	2	1	3	15
2	2	3	1	1	2	9
3	2	2	0	1	0	5
Totals	8	10	3	3	5	29

(6)

70   Teacher's notes

(b) *slug pellet traps:*

	No. slugs trapped					
Day	Bait 1 g	5 g	10 g	15 g	20 g	Total catch
1	8	10	10	9	9	46
2	9	9	8	7	7	40
3	7	5	4	10	6	32
Totals	24	24	22	26	22	118

(6)

2 Slug pellets trap more slugs than carpet/newspaper traps.
Most slugs were trapped on the first morning after traps were laid. Thereafter, the effectiveness of both types of trap decreased. The application of 1 g slug pellets/m² soil surface was as effective in trapping slugs as applications of larger quantities. (3)

3 *Criticisms*
Results were based on a single trial.
The effects of applying pellets at less than 1 g/m² were not investigated.

Improvements
Duplicate all trials.
Extend trials to include the effects of 0.25, 0.5 and 0.75 g slug pellets/m². (3)

4 1. Blue-coloured pellets are easily seen by humans.
2. Coloured pellets are less likely to be eaten by birds and small mammals, which are not attracted by blue-coloured objects. (2)

46 Soil conditioning

Investigation: 40–60 mins

MATERIALS

- 70 g dry sand
- 70 g dry clay
- 10 g Acta-bacta²
- 2 100 cm³ beakers, graduated in 20 cm³ (or 25 cm¹) units
- 50 cm³ (or 100 cm³) measuring cylinder
- filter funnel
- teaspoon
- 4 filter papers
- stop-clock, or watch with a second hand

Instructions and notes

1 Silver sand and builder's sand are suitable for this investigation.

2 Clay can be dug during favourable weather conditions and stored in polythene sacks. After drying, the clay should be ground to a fine powder, using a pestle and mortar, before it is distributed to pupils.

Teacher's notes 71

Specimen answers

1

Time (s)	Vol. water collected in measuring cylinder (cm³)			
	Sand	**Sand + Acta-bacta**	**Clay**	**Clay + Acta-bacta**
0	0	0	0	0
30	22	16	10	15
60	33	25	16	18
90	38	31	23	23
120	41	35	27	26
150	44	39	30	26
180	45	40	32	27
210		40	32	
240		41	32	
270		42	33	

(6)

2 (a) 5 cm³   (b) 8 cm³   (c) 17 cm³
   (d) 23 cm³   (4)

3 (a) $\frac{5}{50} \times 100 = 10\%$ (b) $\frac{8}{50} \times 100 = 16\%$
   (c) $\frac{17}{50} \times 100 = 34\%$ (d) $\frac{23}{50} \times 100 = 46\%$
                                                      (4)

4 (a)

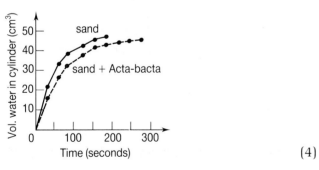

(4)

(b)

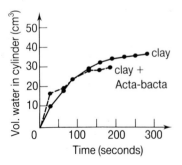

(4)

5 (a) Acta-bacta increases water retention and slows the rate at which water drains through sand. (1)
(b) Acta-bacta increases water retention, but speeds up the rate at which surplus water drains through clay. (1)

6 *Faults*
1. Amounts of sand/clay used were approximate (e.g. 20 cm³ or 25 cm³).
2. Addition of Atca-bacta to sand/clay increased its volume.
3. Conclusions were based on a single set of results.

Improvements
1. Use identical masses (e.g. 50 g) sand/clay.
2. Add Acta-bacta to sand/clay, then weigh out 50 g of the mixture.
3. Obtain at least two sets of results; calculate mean values. (6)

47 Composting

Preparation: 20–30 mins
Investigation: 30–40 mins, over 5–10 days
April–September

Preparation

MATERIALS

- grass mowings (freshly cut)
- large freezer bag (2–3 kg capacity)
- polythene carrier bag (10–15 kg capacity)
- 2 polythene sacks (50 kg capacity)
- 1 kg sucrose
- commercial compost-maker (e.g. Garotta)
- thermometer
- string
- garden fork
- watering can giving a fine spray

Investigation

MATERIALS

- containers filled with grass mowings

Instructions and notes

1 Pupils will need access to large quantities of freshly cut lawn mowings, often readily available near council parks, cricket pitches, tennis courts and bowling greens.

2 Three polythene/plastic containers of different sizes are required. It is suggested a large freezer bag, carrier bag and sacks could be used.

3 Garotta, available from garden centres and hardware stores, is widely used as a compost-maker.

Specimen answers

1

Day	Temperature (°C)				
	Air	2–3 kg container	10–15 kg container	50 kg container	50 kg + sucrose etc. container
1	18	19	18	18	18
2	18	19	19	21	23
3	20	22	23	26	31
4	17	18	19	29	34
5	21	22	23	29	36

Table continued overleaf

Teacher's notes 73

Day	Temperature (°C)				
	Air	2–3 kg container	10–15 kg container	50 kg container	50 kg + sucrose etc. container
6	23	24	26	30	38
7	18	19	23	27	31
8	19	21	23	25	29
9	20	21	22	24	26

(4)

2

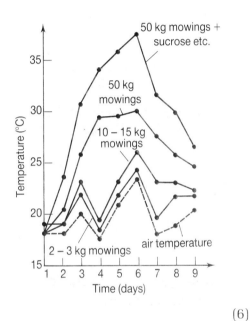

(6)

(6)

3 Grass mowings in the 2–3 kg and 10–15 kg containers failed to generate enough heat to raise the internal temperature at the centre of the containers more than a degree or two above air temperature. Conversely, grass mowings in the sacks generated much more heat as a result of bacterial activity. Heat production was increased by adding sucrose (an activator) and water. The results illustrate the point that the breakdown of plant material is likely to be more rapid in a large compost heap than in a small one due to increased heat production and retention. (3)

4 (a) Mowings in the freezer bag were green and intact, while those in the sacks were brown and beginning to break down.
(b) Higher temperatures in the sacks increased the rate of composting. The primary effect was on enzymes produced by saprophytic bacteria and fungi. (2)

48 Making a biogas generator

Preparation: 60–80 mins

Preparation

MATERIALS

- 2–3 kg kitchen waste
- 2–3 dm³ rain water
- 5 dm³ plastic container
- 500 cm³ gas jar
- beehive shelf
- rubber tubing, 1.5–2.0 m × 1.5–2.5 cm diameter

- plastic bucket
- kitchen mincer or blender
- bradawl (wooden handled)
- paintbrush
- aluminium paint
- Plastic Padding
- Bunsen burner
- scalpel
- eye protection
- access to fume cupboard

Instructions and notes

1 Heat the plastic container only in a fume cupboard. Smoke from the plastic may contain toxic gases.

2 A note on the use of Plastic Padding is given on page 42. This material may not stick to plastic unless the surface has previously been coated with a paint containing powdered aluminium. Once this paint layer is thoroughly dry, Plastic Padding will normally bond to it, forming a water-tight seal.

3 Kitchen waste, consisting mostly of vegetable peelings, leaves, etc., is a suitable raw material. The generation of biogas will be speeded up if this material is finely divided, by passing it through a kitchen mincer or blender. Even when the material is finely divided, it may take several weeks before any biogas is generated.

4 Owing to the production of some hydrogen sulphide, a toxic gas, the generator is best set up out of doors, in a shed or outhouse.

5 Do not attempt to demonstrate the combustible nature of biogas with a glowing splint. Methane/air mixtures are explosive.

Specimen answers

1 Animal manure; straw; remains of farm crops (e.g. cabbage stalks); whey; kitchen waste. (2)

2 Many biogas generators are unsightly. Placing the generator below ground hides it from view, and also reduces the chance of personal injury from fires, explosions, etc. (2)

3 Gases emitted from the biogas generator are obnoxious and toxic. In addition, the gases are inflammable and explosive if mixed with air. Building a biogas generator close to a house would constitute a fire hazard. Furthermore, slurry in the generator may contain pathogenic organisms that could cause disease. (3)

4 Many parts of China and India do not have gas or electricity supplies. A biogas generator provides cheap fuel for heating, cooking and lighting in the home. It also provides a storage tank and a method of disposing of domestic and animal wastes, with the added advantage that slurry from the generator can be used as a fertiliser. (3)

Teacher's notes 75

49 The effects of acid rain

Preparation: *15–20 mins*
Investigation: *30–50 mins, over 4–5 days*

Preparation

MATERIALS

- mustard seeds
- 10 cm³ 0.1 M nitric acid in a suitable container
- 10 cm³ 0.1 M sulphuric acid in a suitable container
- 100 cm³ distilled water in a beaker
- 9 petri dishes
- 6 flat-bottomed tubes
- 9 filter papers
- 2 10 cm³ plastic syringes
- scissors
- glass-marking pen

Investigation

MATERIALS

- petri dishes containing germinating mustard seeds

Instructions and notes

1 Each pupil requires 10 cm³ nitric and 10 cm³ sulphuric acids. These may be supplied in labelled flat-bottomed tubes.

2 It is suggested that the investigation is stopped after the radicles in the dish with water have reached a length of 3–5 cm. This, however, is something that must be left to your own discretion. In order to produce material suitable for class use, one of the following procedures can be adopted. Growth rates can be speeded up by transferring the petri dishes to an incubator at 25°C, or slowed down by transferring them to a refrigerator.

Specimen answers

1

Acid	Molarity	Mean radicle length (cm)
HNO₃	0.0 (water)	4.2
	0.0001	5.2
	0.001	5.1
	0.01	3.8
	0.1	0.3

Table continued overleaf

76   Teacher's notes

Acid	Molarity	Mean radicle length (cm)
H_2SO_4	0.0 (water)	4.2
	0.0001	3.9
	0.001	3.5
	0.01	1.5
	0.1	0.0

(4)

2

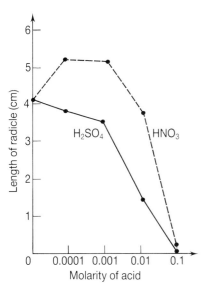

(6)

3 **(a)** The general effect of acid was inhibitory. Concentrated acid was more inhibitory than dilute acid. (1)
(b) Sulphuric acid inhibited germination at all concentrations from 0.0001 to 0.01 M. Nitric acid, however, promoted germination at concentrations of 0.001 M and 0.001 M. (2)

4 Sulphuric acid is highly toxic. Nitric acid is less toxic, and at concentrations below 0.001 M may supply nitrogen to the germinating seedlings, thereby increasing their rate of growth. (2)

50 Water purification

Investigation: 40–60 mins

Investigation

MATERIALS

- 10 g clay
- 10 g sand
- 10 g gravel
- 2 20 cm^3 plastic syringes
- 10 cm^3 plastic syringe
- 2 retort stands, bosses and clamps
- 2 10 cm^3 measuring cylinders
- teaspoon
- stop-clock, or watch with a second hand
- glass-marking pen

Instructions and notes

1 Silver sand and washed gravel (for use in fish tanks) are available from garden centres and pet stores. Clay can be dug from sub-soil, dried in a greenhouse, and powdered in a mortar.

2 Dry clay, sand and gravel can be supplied to pupils in suitable containers, such as petri dishes.

Specimen answers

1

| | Vol. water collected in measuring cylinders (cm³) ||
Time (mins)	Syringe A	Syringe B
0	0.0	0.0
5	2.7	1.6
10	6.2	3.0
15	8.1	4.2
20	8.7	5.4
25	9.1	6.1
30		7.1
35		8.0
40		9.0

(5)

2

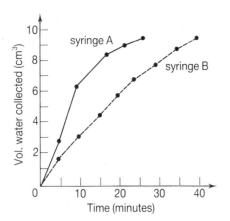

(5)

3 (a) syringe B (1)
(b) syringe A (1)

4 (a) Layering the particles with largest (gravel) at the top and smallest (clay) at the bottom produces the most effective filter bed. Although the rate of filtration is slower, the filtrate does not contain fine particles of mud because they cannot pass through the small pores between small particles at the bottom of the filter bed. (4)

(b) Examining the filtrate beneath a microscope would show if any bacteria, algae and protozoa were still present in the water. (2)

(c) The water would be fit for drinking after sterilisation, either by boiling or by adding chlorine (chlorination) to kill off any remaining micro-organisms. (2)

Further reading

Bentley, G I (1981). *Microbiology: Technicians' Guide*. A.S.E., Hatfield.

Bickerstaff, G E (1987). *Enzymes in Industry and Medicine*. New Studies in Biology. Hodder and Stoughton, Sevenoaks.

Bishop, O N (1984). *Adventures with Microorganisms*. John Murray, London.

Bravery, H E (1976). *Home Booze*. Book Club Associates, London.

British Earthworm Technology Ltd. (1984). International Conference on Earthworms in Waste and Environmental Management: Abstracts.

Department of Education and Science, Welsh Office (1986). *Science 5–16: a Statement of Policy*. HMSO.

Department of Education and Science (1985). *Microbiology an HMI Guide for Schools and Non-advanced Further Education*. HMSO.

Edwards, C A and Lofty, J R (1972). *Biology of Earthworms*. Bookworm Publishing Co., Bungay, Suffolk.

Freeland, P W (1972). GA enhanced amylase synthesis in halved grains of barley (*Hordeum vulgare*): a simple laboratory demonstration. *Journal of Biological Education*, **6**, 369–375.

Freeland, P W (1973). Some applications of glucose-sensitive reagent strips in biology teaching. *School Science Review*, **55**, 190, 14–22.

Freeland, P W (1975). Fingerprint analysis. *Journal of Biological Education*, **9**, (5), 195–200.

Freeland, P W (1975). Some applications of agar-gel diffusion techniques. *School Science Review*, **56**, 195, 274–287.

Freeland, P W (1985). *Problems in Practical Advanced Level Biology*. Hodder and Stoughton, Sevenoaks.

Freeland, P W (1987). *Investigations for GCSE Biology*. Hodder and Stoughton, Sevenoaks.

Fry, P J (1977). *Micro-organisms*. Schools Council. Hodder and Stoughton, Sevenoaks.

Glynn, C (1977). *Cheese and Cheese Making*. Macdonald, London.

Hammond, S M and Lambert, P A (1978). *Antibiotics and Antimicrobial Activity*. Studies in Biology No. 90. Edward Arnold, London.

Higgins, I J, Best D J and Jones, J (1985). *Biotechnology*. Blackwell, Oxford.

Inchley, C J (1981). *Immunobiology*. Studies in Biology No. 128. Edward Arnold, London.

James, T H (1987). *Forensic Science*. Stanley Thornes, Cheltenham.

Mantell, S M, Matthews, J A and McKee, R A (1985). *Principles of Plant Biotechnology*. Blackwell, Oxford.

Satelle, D B (1988). *Biotechnology in Perspective*. Hobsons, Cambridge.

Satelle, D B and Katz, J (1988). *Biotechnology in Focus*. Hobsons, Cambridge.

Sharp, J A (1977). *An Introduction to Animal Tissue Culture*. Studies in Biology. Edward Arnold, London.

Smith, J E (1988). *Biotechnology*. New Studies in Biology. Hodder and Stoughton, Sevenoaks.

Teasdale, J (1987). *Biotechnology*. Stanley Thornes, Cheltenham.

Terry, J (1987). What fermenter? *Journal of Biological Education*. **21** (3), 163–166.

Thear, K (1983). *Home Dairying*. Batsford, London.

Wallwork, J A (1983). *Earthworm Biology*. Studies in Biology No. 161. Edward Arnold, London.

Warr, J R (1984). *Genetic Engineering in Higher Organisms*. Studies in Biology No. 161. Edward Arnold, London.

Wymer, P E O (1987). *Practical Microbiology and Biotechnology for Schools*. Macdonald Education, London.

Names and addresses of suppliers

1 BDH Chemicals Ltd.,
  Poole,
  Dorset.
  BH12 4NN

2 Easi-Gro Ltd.,
  Marlborough Road,
  Aldbourne,
  Wiltshire.
  SN8 2DD

3 Griffin and George Ltd.,
  Gerrard Biological Centre,
  Worthing Road,
  East Preston,
  West Sussex.
  BN16 1AS

4 Philip Harris Biological Ltd.,
  Oldmixon,
  Weston-super-Mare,
  Avon.
  BS24 9BJ

5 Hughes and Hughes (Enzymes) Ltd.,
  Elms Industrial Estate,
  Church Road,
  Harold Wood,
  Romford,
  Essex.
  RM3 0HR

Unfamiliar items in the lists of materials are marked by a number indicating the supplier (e.g. Acta-bacta[2]).

PUPIL'S SHEETS

INVESTIGATIONS IN Applied Biology and Biotechnology

Peter Freeland

Contents: Pupil's Sheets

1 Enzyme isolation 5
2 Enzymes in washing-powders 8
3 The best temperature for a washing-powder 11
4 Tenderising meat 14
5 Peas and beans (pulses) 16
6 An enzyme inhibitor in beans 19
7 Protein precipitation 22
8 Solvents in food manufacture 25
9 Particle size in cereal products 27
10 Physical properties of milk 29
11 When does custard thicken? 32
12 Making and using yogurt 34
13 Cheese-making 36
14 Cheese tasting 39
15 Acid and sugar in citrus fruits 41
16 Making sweeter sugar 44
17 Energy content of foods 47
18 'E' is for artificial colouring 49
19 Natural colouring 51
20 Research with extracted chloroplasts 53
21 Bread-making 55
22 Wine-making 58
23 Malting barley 61
24 Alcohol production 65
25 Making a fermenter 68

26 Cell counts 71
27 Which antibiotic? 73
28 Daily urine output 76
29 Drinking and driving 78
30 Variations in a human population 80
31 Fingerprint analysis 82
32 How exercise affects your heart 84
33 Heart and breathing rates during exercise 86
34 Lung capacity 88
35 Measuring grip strength 91
36 Obesity 93
37 Successful seed germination 95
38 New bulbs from old 97
39 New plants from cuttings 100
40 Rooting cuttings in a gel 103
41 Micropropagation 106
42 Pollination 109
43 Longer life for cut flowers 112
44 Dispersing weed seeds 114
45 Trapping and killing slugs 116
46 Soil conditioning 119
47 Composting 121
48 Making a biogas generator 123
49 The effects of acid rain 125
50 Water purification 127

1 Enzyme isolation

Preparation: Part 1 – 5 mins
Part 2 – 20–30 mins
(after 2–3 days)
Investigation: 30–40 mins

Enzymes are **catalysts** produced by living cells. Some cells produce extracellular enzymes, which are released into their surroundings. Others produce intracellular enzymes, which are used inside the cells that produce them. The demand for enzymes is growing all the time and the extraction and purification of enzymes, called **isolation**, is an important commercial process. Isolating extracellular enzymes is fairly straightforward. On the other hand, the isolation of intracellular enzymes is more of a challenge because the cells must be crushed, and the enzyme separated out from all the other cell contents.

Amylase is an enzyme produced by germinating pea and bean seeds. It converts starch into the disaccharide, maltose:

$$\text{starch} \underset{}{\overset{\text{amylase}}{\rightleftharpoons}} \text{maltose}$$

In this investigation you will find out if germinating seeds produce amylase as an extracellular or intracellular enzyme. The investigation will take you through the first three stages of enzyme isolation:

production → **extraction** → **concentration**
(in seeds)   (by filtration)   (by removal of water)

It is unlikely that there will be enough enzyme for the fourth stage, which is **purification** by precipitation and recrystallisation, but this is what would happen if the enzyme was to be used, for example, in biological washing-powder.

Preparation

PART 1

Materials

- 100 g pea or bean seeds
- 500 cm³ beaker

Method

Put the pea or bean seeds into the beaker and add 200 cm³ of water. Leave the seeds in the water for 2–3 days.

PART 2

Materials

- soaked pea or bean seeds in 500 cm³ beaker
- starch-agar plate
- 100 cm³ beaker
- 100 cm³ measuring cylinder
- 5 test tubes fitted with rubber bungs
- 5 1 cm³ plastic syringes
- filter funnel
- no. 6 cork borer
- kitchen mixer

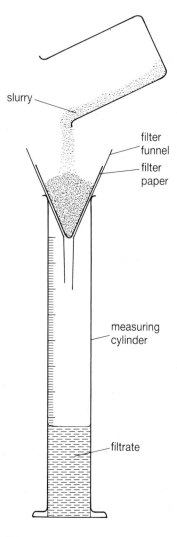

Fig 1

- waterbath maintained at 30°C
- incubator maintained at 25°C
- filter papers
- adhesive tape
- glass-marking pen
- access to refrigerator

Method

1 Pour off about 5 cm³ of the liquid surrounding the seeds into a test tube. Label this 'extracellular fluid'. Replace the rubber bung and store the tube in a refrigerator.

2 Transfer the seeds, and the liquid surrounding them, to the kitchen mixer. Grind the seeds for 2–3 minutes, then tip the slurry back into the beaker.

3 Place the filter funnel, lined with filter paper, in the top of the measuring cylinder (see Fig 1). Filter the slurry and keep the filtrate. Pour about 5 cm³ of the filtrate into a test tube and label this 'intracellular + extracellular fluid'. Replace the rubber bung and store the tube in a refrigerator.

4 Pour 100 cm³ of filtrate into the 100 cm³ beaker. Transfer the beaker to a waterbath maintained at 30°C. Observe the level of liquid in the beaker as the water evaporates. Draw off 5 cm³ of filtrate when the water level reaches 75, 50 and 25 cm³. Put each sample into a labelled test tube and store the tubes in a refrigerator.

5 Take the starch-agar plate and cut five holes in it, spaced out as shown in Fig 2. Use the plastic syringes to transfer the extracellular fluid and the four different concentrations of filtrate to the holes. Put four drops of liquid into each hole. Seal the agar plate with adhesive tape and number the holes.

6 Transfer the starch-agar plate to an incubator at 25°C, and incubate for 1–3 hours.

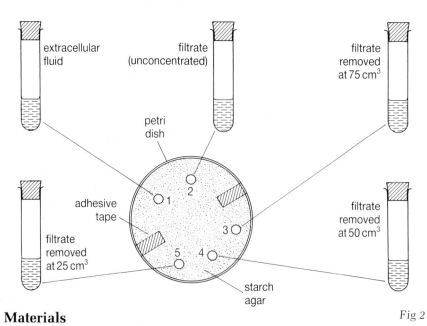

Fig 2

Investigation

Materials

- starch-agar plate containing seed extracts
- iodine solution
- ruler, graduated in mm

Method

1 Remove the agar plate from the incubator.
2 Pour iodine solution over the starch-agar. Wait until all the starch has stained (blue-black).
3 After staining, measure and record the diameter of each clear (starch-free) zone that has developed around the holes.

QUESTIONS

1 **(a)** Copy and complete the table. (The area of the clear zone is calculated from the formula πr^2, where r = radius and π = 3.14.)   (5)

Hole no.	Diameter of clear zone (cm)	Area of clear zone (cm^2)
1		
2		
3		
4		
5		

(b) Was (i) extracellular and (ii) intracellular amylase produced by the germinating seeds?   (2)

2 If the concentration of amylase in hole 2 is 1.0 (arbitrary units), what is the concentration of amylase in
   (a) hole 3
   (b) hole 4
   (c) hole 5?
Show how you arrived at your answers.   (3)
   (d) Plot a graph of the area of the clear zone against the concentration of enzyme in holes 2–4.   (4)

(e) What do you conclude about the concentration of enzyme and the area of gel cleared of starch?   (2)

3 If both extracellular and intracellular enzyme is produced, how could you calculate the percentage of
   (a) extracellular
   (b) intracellular
enzyme in the filtrate? Your answer should include at least one equation.   (4)

Taking it further

1 Carry out a similar investigation of the protease (a protein-digesting enzyme) produced by germinating seeds. Use an agar plate containing 'Marvel' milk (2%$^{w/v}$) as substrate for any protease that might be produced. Similarly, try to extract lipase (a fat-digesting enzyme) from germinating sunflower seeds. Use a salad cream (1%$^{w/v}$) agar for testing the extracellular fluid and filtrate for lipase.

2 Amylase, lipase and protease are produced by some bacteria, which release these enzymes into their surroundings. Find out more information about bacteria, and the enzymes that they release. Draw up a table of not more than 10 species with the extracellular enzymes that they produce.

3 Design and draw a small-scale plant that could be used for the continuous production of extracellular enzymes, using either bacteria or other micro-organisms.

Investigations in Applied Biology and Biotechnology © 1990 Peter Freeland. Published by Hodder & Stoughton

2 Enzymes in washing-powders

Preparation (day 1): *20–30 mins*
Investigation (day 2): *30–40 mins*

All household washing-powders contain soaps or detergents, which reduce the surface tension of water. This loosens surface dirt so that it can be washed away from clothes more easily. The so-called **biological washing-powders** also contain enzymes, usually extracted from bacteria. These enzymes break down large molecules, such as starch and proteins, into smaller ones that will dissolve in water. Stubborn stains on clothing, caused by gravy or blood, wash off more easily after the enzymes have acted on them.

In this investigation there are three main aims:

(a) To identify the enzymes present in two biological washing-powders.
(b) To compare the levels of enzyme activity in the washing-powders.
(c) To find out how much enzyme activity has been lost by an old packet of washing-powder.

You are provided with agar plates containing (a) starch (for finding out about amylase activity), (b) milk (for protease activity) and (c) mayonnaise (for lipase activity).

Preparation

Materials

- 5 cm³ 1%$^{w/v}$ solution of powder A in a labelled container
- 5 cm³ 1%$^{w/v}$ solution of powder A$_1$ in a labelled container
- 5 cm³ 1%$^{w/v}$ solution of powder B in a labelled container
- starch-agar plate
- milk-agar plate
- mayonnaise-agar plate
- 1 cm³ plastic syringe
- no. 6 cork borer
- adhesive tape
- glass-marking pen
- incubator maintained at 25°C

Method

You have three solutions of different washing-powders, labelled A, A$_1$ and B. Powder A$_1$ is the same product as powder A, but is older (it was bought at least 6 months ago).

1 Use the cork borer to cut three holes in each agar plate, spaced out as shown in Fig 1.

2 Take the starch-agar plate. Use the syringe to put four drops of solution A into hole 1. Similarly, put four drops of solution A$_1$ into hole 2, and four drops of solution B into hole 3.

3 Take the milk-agar plate, then the mayonnaise-agar plate, and repeat the procedure outlined above, with three holes per plate.

4 Replace the lids of the petri dishes. Seal them with adhesive tape. Label the contents of each hole.

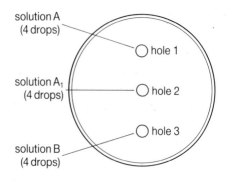

Fig 1

Investigations in Applied Biology and Biotechnology © 1990 Peter Freeland. Published by Hodder & Stoughton

5 Put the petri dishes into an incubator at 25°C. Allow 12−24 hours for the enzymes in the solutions to diffuse into the agar.

Investigation

Materials

- incubated starch-agar plate
- incubated milk-agar plate
- incubated mayonnaise-agar plate
- iodine solution
- ruler, graduated in mm

Method

1 Pour the iodine solution over the starch-agar plate. Pour away the iodine solution after 3−5 minutes.

2 Measure and record the diameter of any clear (transluscent) zones that have developed around the holes in each petri dish.

3 Similarly, measure and record the diameter of any clear zone that has developed around a hole in the milk-agar and mayonnaise-agar plates.

QUESTIONS

1 Copy and complete the table. (The area of the clear zone around a hole can be calculated from the formula πr^2, where r = radius and $\pi = 3.14$.) (9)

Agar plate	Solution	Diameter of clear zone (cm)	Area of clear zone (cm²)
starch	A		
	A_1		
	B		
milk	A		
	A_1		
	B		
mayonnaise	A		
	A_1		
	B		

2 Which enzymes (amylase, protease, lipase) were present in each of the solutions? (6)

3 Which powder, A or B, was most active? Give a reason for your answer. (2)

4 (a) What was the effect of keeping powder A_1 for a long time before it was used?
(b) Calculate the percentage increase/decrease in activity of one enzyme in this powder. Show how you arrived at your answer. (3)

Investigations in Applied Biology and Biotechnology © 1990 Peter Freeland. Published by Hodder & Stoughton

Taking it further

1 Make a list of all the biological washing-powders stocked by your local supermarket. Try to find out which one is (a) the most active and (b) the least active in terms of enzyme activity.

2 Using gravy-stained pieces of muslin, cotton or linen, try to find out how long it takes for the stain to be removed by a 5%$^{w/v}$ solution of the washing-powder at room temperature.

3 The best temperature for a washing-powder

Preparation: 50–60 mins
Investigation: 30–40 mins

As you may have seen in the last investigation, biological washing-powders contain enzymes. These enzymes, usually **amylases** and **proteases**, help to break down large organic molecules into smaller ones which are more easily washed out of clothes. While it is usual to boil clothes in non-biological washing powders, powders that contain enzymes work best at lower temperatures. The optimal temperature for most enzymes lies within the range 30–35°C. There are, however, some exceptions. For example, bacteria living in hot springs have enzymes with an optimal temperature 10°C or more higher.

In this investigation you will find out how well the protease in a biological washing-powder, and **trypsin** (another protease) from a cow, work at different temperatures. What is the optimal temperature of each enzyme? Is the protease in the washing-powder denatured at the same rate as trypsin, or more slowly at higher temperatures?

Precautions

Take care when handling hot syringes.

Preparation

Materials

- 5 cm^3 1%$^{w/v}$ trypsin solution in a 5 cm^3 plastic syringe
- 5 cm^3 1%$^{w/v}$ biological washing-powder solution in a 5 cm^3 plastic syringe
- 2 milk-agar plates
- 250 cm^3 beaker
- thermometer
- no. 6 cork borer
- adhesive tape
- Bunsen burner, tripod and gauze
- paper towel
- incubator maintained at 25°C
- glass-marking pen
- eye protection

Method

1 Use the cork borer to cut seven holes in each agar plate, spaced out and arranged as shown in Fig 1. Label the dishes 'trypsin' and 'washing-powder'.

2 Use the marking pen to label the solutions in the syringes.

3 Put on your eye protection. Light the Bunsen burner. Turn down the gas, to give a small, steady flame.

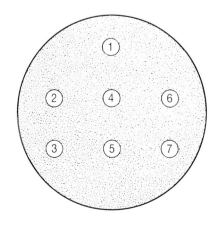

Fig 1

4 Half fill the 250 cm³ beaker with water, and put it on the gauze above the tripod. Put the syringes and thermometer into the beaker, as shown in Fig 2. (The top of the syringe barrels should be above the water level.)

5 Heat the water in the beaker to 30°C, stirring with the thermometer. Keep the water at 30°C for 3 minutes. (You may have to remove the Bunsen burner for a time.)

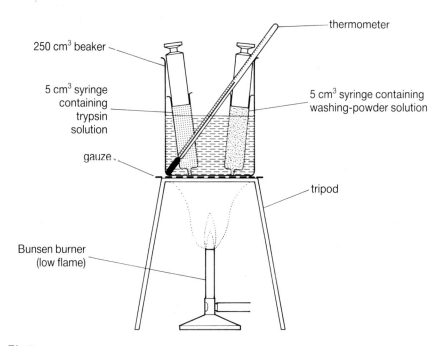

Fig 2

6 Carefully remove the syringes. Use the paper towel to dry them on the outside. As quickly as possible, put five drops of trypsin solution into hole 1 of the 'trypsin' dish. Similarly, put five drops of the washing-powder solution into the 'washing-powder' dish. Return the syringes to the beaker.

7 Repeat instructions 5 and 6 for each of the following temperatures: 40°C, 50°C, 60°C, 70°C, 80°C, 90°C.

8 Put the lids on the petri dishes. Seal them with adhesive tape and number the holes. Incubate the dishes at 25°C for 12–24 hours.

Investigation

Materials

- 2 incubated milk-agar plates
- ruler, graduated in mm

Method

Measure and record the diameter of the clear (transluscent) zone around each hole.

QUESTIONS

1 Copy and complete the table. (Calculate the area of the clear zone from the formula πr^2, where r = radius and π = 3.14.) (14)

Hole no.	Temperature (°C)	Diameter of clear zone (cm)		Area of clear zone (cm^2)	
		Trypsin	Washing-powder	Trypsin	Washing-powder
1	30				
2	40				
3	50				
4	60				
5	70				
6	80				
7	90				

2 Draw a graph to compare protease activity at different temperatures in the two solutions. (5)

3 Calculate the percentage decrease in
  (a) trypsin acvitity
  (b) protease activity
in washing-powder at 60°C. Show how you arrived at your answers. (4)

4 What do you conclude from the investigation? (4)

5 Give three criticisms of the method. (3)

1 Design, write and carry out an investigation of the effect of temperature on amylase activity using amylase from (a) a plant or animal source and (b) a washing-powder. How would you find out if the two amylases worked best at the same, or different, hydrogen ion concentrations (pH)?

4 Tenderising meat

Preparation: 20–30 mins
Investigation: 30–40 mins

Some cuts of meat, especially beef and pork, can become tough, dry and chewy after cooking. Young children, or elderly people who have lost their natural teeth, may find it difficult to chew and swallow meat that is tough and fibrous. Cooks have known for many years that meat can be made tender (tenderised) by soaking the raw meat in natural acids, such as lemon juice or vinegar. This treatment, usually set up the evening before the meat is cooked, is called **marinating**. Scientists now know that marination causes the muscle fibres in meat to absorb more water so that they remain tender and succulent during cooking.

In this investigation you will find out how pH affects the ability of raw meat to absorb water. You have some diced steak, and buffer solutions at different pH. Carry out the instructions to find out how the different treatments affect the meat.

Preparation

Materials

- steak (400–500 g)
- 6 bottles, each containing 100 cm^3 of a buffer solution (e.g. pH 4.0, 6.8, 7.0, 8.0, 9.0 and 9.2)
- 6 100 cm^3 beakers
- forceps
- scalpel
- glass-marking pen
- Clingfilm
- top-pan balance

Method

1 Weigh each beaker. Write the mass (g) of each beaker on the outside (see Fig 1).

2 Cut the steak into small cubes, each about 1 cm^3. Put 10–15 pieces of cubed steak into each beaker.

3 Re-weigh the beakers. Write the mass (g) of the raw meat on the outside of each beaker.

4 Label each beaker with the pH of one of the buffer solutions. Add the correct buffer to each beaker, making sure that the meat is covered.

5 Cover each beaker with Clingfilm. Leave the beakers in a cool place (10–15°C) for 12–24 hours.

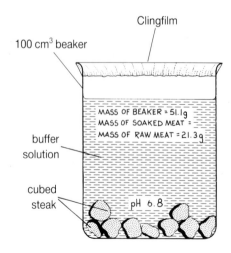

Fig 1

Investigation

Materials

- 6 100 cm³ beakers containing cubed steak in buffer solutions
- forceps
- paper tissues
- top-pan balance

Method

1 Pour away the buffer solutions from each beaker. Take each beaker in turn. Use forceps to remove the cubed steak. Dry it on the paper tissues.

2 Weigh the meat from each beaker. Write the mass (g) of the meat on the outside of the beaker.

3 Record the colour of the soaked meat in each beaker.

QUESTIONS

1 Copy and complete the table. (6)

pH of buffer	Mass of meat (g) Before soaking	Mass of meat (g) After soaking	Increase (+)/ decrease (−) (g)	% increase/ decrease
4.0				
6.8				
7.0				
8.0				
9.0				
9.2				

2 Plot your results as a graph. (4)

3 What do you conclude about the effect of pH on
  (a) water absorption (2)
  (b) colour of the meat after soaking? (2)

4 Give three criticisms of the method and suggest three improvements. (6)

Taking it further

1 Try to find out how much water meat absorbs when it is marinated in (a) lemon juice and (b) vinegar. How does marination affect (i) the flavour and (ii) the texture of meat?

2 What is the importance of your results to producers of canned meats? How should they treat meat to ensure that their product is always succulent and tender?

3 Does marination have any disadvantages? If so, what are they and how might they be overcome?

5 Peas and beans (pulses)

Preparation: 10–15 mins
Investigation: 60–80 mins

Peas and beans belong to a group of plants known as **legumes**. The edible seeds of legumes, called **pulses**, are a rich source of carbohydrates, protein and fibre. Supermarkets usually stock several different types, from different parts of the world, which you can identify by using the key on Sheet 18. Seed importers may want to know if a particular batch of seed is freshly harvested or old. They can tell by testing the seeds for enzyme activity. Freshly harvested seeds contain active enzymes. Old seeds show much less enzyme activity, while dead ones, unfit for human consumption, have none.

The table below shows some of the tests you will carry out on packeted peas and beans to find out if they are alive or dead. People who sell seeds for sowing can use the same tests to find out if their seeds are still alive, or **viable**, without waiting to see if they germinate (see Investigation 37).

Precautions

Wear eye protection and plastic gloves when handling hydrogen peroxide.

Condition of seed	Property	Chemical test
living	release glucose when soaked	(a) turns Clinistix from pink → blue (b) forms red precipitate when boiled with Benedict's solution
	release carbon dioxide when soaked in water	decolourises an alkaline solution, coloured pink with phenolphthalein
	release bubbles of oxygen when hydrogen peroxide is added to crushed samples	add a few drops of 10-vol hydrogen peroxide to crushed seeds; look for bubbles
dead	no glucose, carbon dioxide or oxygen produced in any of the tests	/////////

Preparation

Materials

- 6 different types of pea/bean
- 6 boiling tubes in a rack
- 65 cm³ 0.01 M NaOH solution, coloured pink with phenolphthalein (indicator solution)
- 10 cm³ plastic syringe
- ruler, graduated in mm
- glass-marking pen

Method

1 Label the boiling tubes A–F. Draw a horizontal line on each tube at 2.5 cm from the base. Put one type of dry pea/bean seeds into each tube to the 2.5 cm mark.

2 Use the syringe to add 10 cm³ of indicator solution to each tube.

3 Leave the tubes standing on the bench overnight. Examine them the next day.

Investigation

Materials

- 6 boiling tubes containing soaked seeds
- 6 (or more) different types of dry seeds (numbered)
- 6 Clinistix reagent strips
- dropping bottle containing 10-vol H_2O_2 solution
- 6 text tubes in a rack
- white tile
- scalpel
- ruler, graduated in mm
- glass-marking pen
- plastic gloves
- eye protection

Method

1 Look at the peas/beans that were soaked in the indicator solution. In which tubes have the seeds decolourised the solution? Record your results.

2 Dip a Clinistix reagent strip into each solution. In which tubes is there enough glucose to turn the reagent strip from pink to blue? Record your results.

3 Put on your plastic gloves and eye protection. Label the test tubes A–F.

4 Crush one or more seeds from each boiling tube on the white tile. Put the crushed seeds from each boiling tube into a test tube labelled with the same letter. Add about 10 drops of hydrogen peroxide solution to each test tube.

5 In which tubes do the crushed seeds produce bubbles of oxygen? Record your results.

QUESTIONS

1 Copy and complete the table, extending it if necessary to include more than six types of seeds. (18)

Test	Seeds					
	A	**B**	**C**	**D**	**E**	**F**
indicator (colourless/pink)						
glucose (present/absent)						
oxygen (present/absent)						

Investigations in Applied Biology and Biotechnology © 1990 Peter Freeland. Published by Hodder & Stoughton

2 What do you conclude about the seeds you have tested? (2)

3 You have six or more dry, numbered peas/beans. Use the following key to identify them. (Measurements refer to the diameter of a typical seed.)

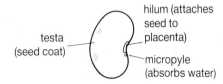

Fig 1 a kidney-shaped bean seed

1. Seeds spherical (peas)   2
   Seeds kidney-shaped (beans) (see Fig 1)   3

2. Testa smooth/wrinkled, green-yellow
      (0.6 cm)   **pea**
   Testa wrinkled, yellow-brown, with
      pointed projection (0.9 cm)   **chick pea**

3. Seed almost spherical, testa yellow-
      brown, shiny (0.6 cm)   **soya bean**
   Seed markedly kidney-shaped   4

4. Hilum (black) at base of seed
      (1.6 cm)   **broad bean**
   Hilum (white) at side of seed   5

5. Seed more than 1.6 cm in length   6
   Seed less than 1.6 cm in length   7

6. Testa white (1.3 cm)   **butter (lima) bean**
   Testa red-brown, speckled black
      (1.2 cm)   **runner bean**

7. Testa of two or more colours   8
   Testa uniformly coloured   9

8. Testa pink-brown, speckled brown-black
      (0.7 cm)   **pinto bean**
   Testa black (around hilum) and white-
      grey (0.6 cm)   **black-eyed bean**

9. Testa red (0.9 cm)   **red kidney bean**
   Testa white (0.7 cm)   **haricot bean**
   Testa brown (0.6 cm)   **dwarf French bean**
   Testa green (0.3 cm)   **Mung bean**

Copy and complete the table. (5)

Number of pea/bean	Sequence of key numbers used in identification	Name of pea/bean
1		
2		
3		
4		
5		
6		

Taking it further

1 Compare the weights and prices of pea/bean seeds sold in the food and garden sections of your local supermarket. If you wanted seeds for sowing, which ones seem to offer the best value for money? Design and carry out an investigation to find out if pea/bean seeds from the food section germinate and grow as well as those from the garden section.

2 Many bean seeds are brightly coloured, especially at the end of the growing season, before the pods have dried. These bright colours serve as a warning to animals.
Bearing in mind that humans and many other animals eat bean seeds, try to find out the significance of the bright warning colours and patterns. Why should humans not eat red kidney beans, black-eyed beans or haricot beans unless they have been soaked in water for several hours and boiled for at least 10 minutes?

6 An enzyme inhibitor in beans

Preparation: *day 1 – 5 mins*
day 2 – 30 mins
Investigation: *30–50 mins*

Supermarkets now carry a wide range of packeted dried beans, including red kidney, black-eyed, haricot and soya. All of these beans are a good source of protein. Indeed, people who are vegetarians may eat them as one of the main sources of protein in their diet. When preparing dried beans to eat, it is important to carry out the instructions on the packet. The packets always carry a warning (see Fig 1). This warning is given because the raw beans contain a poisonous compound, or **toxin**. The toxin is in fact a protein, but one that can cause harm by inhibiting enzymes in the human alimentary canal. Boiling the beans denatures the protein, so they can be eaten without any risk of stomach upsets.

In this investigation you will examine the effects of extracts from red kidney beans, both cooked and uncooked, on the enzyme trypsin. This enzyme, which digests proteins, is produced by the pancreas and is made active in the small intestine.

Fig 1

Preparation

DAY 1

Materials

- 200 g red kidney beans
- 250 cm³ beaker
- household bleach

Method

1 Put the kidney beans into the beaker. Add tap water to the 250 cm³ mark. Add 4–5 drops of household bleach. (This should stop the growth of bacteria and fungi.)

2 Leave the beans standing in the water for at least 24 hours.

DAY 2

Materials

- soaked kidney beans
- milk-agar plate
- 5 cm^3 1%$^{w/v}$ trypsin solution
- 3 100 cm^3 beakers
- 4 1 cm^3 plastic syringes
- no. 6 cork borer
- glass-marking pen
- adhesive tape
- Bunsen burner
- kitchen mixer
- incubator maintained at 25°C
- eye protection

Method

1 Pour about 50 cm^3 of the water surrounding the beans into one of the 100 cm^3 beakers. Label this 'bean water'.

2 Pour about 50 cm^3 of bean water and 25 beans into the kitchen mixer. Grind the beans to a pulp. Pour this into a clean 100 cm^3 beaker. Label it 'raw bean pulp'. Wash out the kitchen mixer.

3 Put on your eye protection. Light the Bunsen burner. Heat the remaining beans and water in the 250 cm^3 beaker over a Bunsen burner until the beans have boiled for at least 10 minutes. Allow the beans to cool. Pour about 50 cm^3 of the boiled bean water and 25 beans into the kitchen mixer. Grind the beans to a pulp. Empty this into a clean beaker, and label it 'boiled bean pulp'.

4 Use the cork borer to cut four holes in the agar place, spaced out as shown in Fig 2.

5 Use a plastic syringe to put two drops of trypsin solution into each hole. Add two drops of clean water to the trypsin in hole 1. Put one of the remaining syringes into each of the beakers containing bean extract. Use each syringe in turn to add two drops of (a) bean water to hole 2, (b) raw bean pulp to hole 3 and (c) boiled bean pulp to hole 4.

6 Replace the lid of the petri dish. Seal it with adhesive tape. Label holes 1–4 on the lid of the dish.

7 Put the petri dish into an incubator at 25°C and let it stand for 12–24 hours.

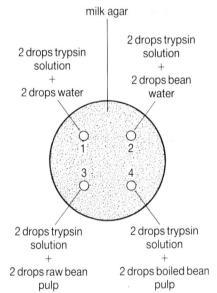

Fig 2

Investigation

Materials

- petri dish containing trypsin and bean extracts
- ruler, graduated in mm

Method

1 Measure and record the diameter of the clear (transluscent) zone surrounding each hole.

QUESTIONS

1 Copy and complete the following table. (The zone containing digested milk can be calculated from the formula πr^2, where r = radius and $\pi = 3.14$.)

Hole no.	Diameter of clear zone (cm)	Area of clear zone (cm)
1		
2		
3		
4		

2 What do you conclude from your results? (5)

3 Explain how errors might have happened if
(a) bleach (or some other antiseptic) had not been added to the water in which the dry kidney beans were soaked (2)
(b) water had not been added to the trypsin solution in hole 1. (2)

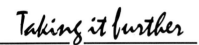

1 You have found out how an inhibitor in beans affects the activity of trypsin. Design, write and carry out an investigation to find out how this substance affects the activity of (a) amylase and (b) pepsin.

2 Write down four possible uses of enzyme inhibitors, assuming that they have no harmful effects on humans.

Protein precipitation

Investigation: 30–40 mins

In the preparation of some foods, enzymes and medicines, you may need to separate proteins from other molecules in solution. One way of extracting a protein is to add a chemical that will cause the protein to be precipitated out of solution. The chemicals used to do this include acids, alkalis, salt, ammonium sulphate, acetate and alcohols, especially ethanol and methanol.

Casein is a white, soluble protein present in milk. It can be extracted from whey, the liquid left over after cheese and butter have been made. Dried, skimmed milk consists largely of casein. Biologists use it in nutrient agar for growing bacteria. It is also used as a binder in paper manufacture to produce a glossy finish suitable for high quality colour printing. You can also make an attractive, but brittle, plastic for buttons, buckles and knitting needles when you pass casein paste into formaldehyde (methanol) solution.

In this investigation you will find out which chemicals precipitate casein most effectively from solution.

Investigation

Materials

- 20 cm^3 soluble casein solution in a beaker
- 20 cm^3 dilute (2 M) hydrochloric acid in a beaker
- 20 cm^3 distilled water in a beaker
- 5 numbered test tubes in a rack, each with a different precipitating agent
- 6 test tubes
- 2 5 cm^3 plastic syringes
- glass-marking pen

Method

1 You are provided with five numbered test tubes, each containing a precipitating agent, as follows:

tube 1: saturated sodium chloride solution
tube 2: dilute (2 M) sulphuric acid
tube 3: ethanol
tube 4: saturated ammonium sulphate solution
tube 5: dilute (2 M) sodium hydroxide solution

Use a 5 cm^3 syringe to add 1 cm^3 of casein solution to each test tube. Use the following symbols to record your results: × = no precipitate; √ = hazy; √√ = cloudy; √√√ = dense crystalline precipitate.

2 Remove the test tubes from the rack and replace them with the six empty test tubes. Number the tubes 1 to 6.

3 Use a clean syringe to put 5 cm^3 of hydrochloric acid into tube 1. In the remaining five tubes mix hydrochloric acid with distilled water (total volume = 5.0 cm^3) to give a range of HCl concentrations.

4 Add 1 cm³ of casein solution to each tube.

5 Record your results using the symbols listed in instruction 1.

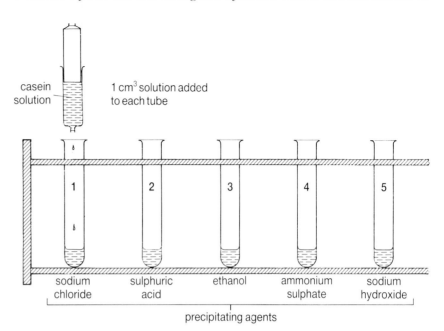

Fig 1

QUESTIONS

1 (a) Copy and complete the table. (5)

Tube no.	Precipitating agent	Amount of precipitate
1	sodium chloride	
2	sulphuric acid	
3	ethanol	
4	ammonium sulphate	
5	sodium hydroxide	

(b) Which compound is most suitable for precipitating casein from solution? (1)

2 (a) Copy and complete the table. (5)

Tube no.	Vol. water (cm³)	Vol. HCl (cm³)	Concentration of HCl	Amount of precipitate
1				
2				
3				
4				
5				

(b) What is the lowest concentration of HCl needed to precipitate casein? (1)

(c) Why is it important to know the lowest concentration of a precipitating agent that will precipitate a protein? (2)

(d) What further treatments would you have to give the casein before you could market it? (3)

Taking it further

1 Some of the precipitating agents that you used in this investigation are used to precipitate enzymes from solution. Which are the most suitable agents for precipitating enzymes? Try to precipitate amylase and protease from a biological washing-powder and soya bean flour, using the chemicals given in the introduction. How will you test for amylase and protease activity?

2 Allow milk to sour, then separate the whey from the curd. Try to precipitate casein from the whey. What is the maximum yield of casein that you can produce from 100 cm^3 whey? Why are maximum yields important in industrial processes?

Solvents in food manufacture

Investigation: 30–45 mins

Milled cereal products, such as porridge oats, consist of particles of many different sizes, ranging from microscopic to about 0.5 cm in length. These products also contain nutrients in the form of sugars, starch and proteins. The particles may be separated and the nutrients extracted using a solvent. Milled cereal is mixed with the solvent, shaken, and left to stand until the **separation** or **extraction** is complete.

Imagine a manufacturer of cereal products wants you to separate milled oats into particles of different sizes. The manufacturer also wants you to extract starch so that the product is less fattening, and proteins so that people who are allergic to cereal proteins are not affected by the product. You are given some porridge oats and four different solvents. Carry out this investigation to find out (a) how effectively the solvents separate fine from coarse particles and (b) which nutrients are extracted by each solvent.

Precautions

Solvent D is an acid – handle it with care.

Investigation

Materials

- 3–5 g porridge oats
- 2 Clinistix reagent strips
- 2 Albustix reagent strips
- 4 test tubes fitted with corks or rubber bungs
- test tube rack
- 20 cm^3 solvent A
- 20 cm^3 solvent B
- 20 cm^3 solvent C
- 20 cm^3 solvent D
- iodine solution
- 1 cm^3 plastic syringe
- 10 cm^3 plastic syringe
- ruler, graduated in mm
- scissors
- glass-marking pen

Method

1 Label the test tubes A, B, C and D. Use the marking pen to draw a horizontal line across each tube at 1.5 cm from the base.

2 Put porridge oats into each tube to a depth of 1.5 cm. Gently tap the tubes and add more oats until the level of the line is reached. Stand your tubes in the test tube rack.

3 Use the 10 cm^3 plastic syringe to add 15 cm^3 of solvent A to tube A. Wash out the syringe and shake it until it is dry. Add 15 cm^3 of each of the other solvents to the tube with the same label. Wash and dry the syringe after each filling.

4 Fit a cork or rubber bung to each tube. Shake each tube vigorously for about 30 seconds, then return it to the rack.

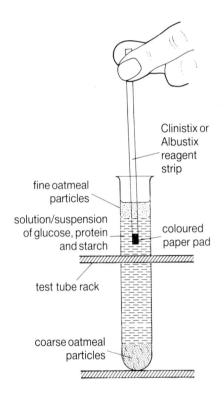

Fig 1

5 Allow 2–3 minutes for the particles to settle. After they have settled, measure and record the depth of particles at (a) the top and (b) the bottom of each tube.

6 Use the scissors to cut the Clinistix and Albustix reagent strips in half lengthways. Do not handle the coloured paper pad at one end of the strip.

7 Take the four Clinistix strips. Dip one into each tube (see Fig 1) then lay the strips on the bench surface beside the tubes they were dipped in.

Clinistix is a test for glucose. If glucose is present in the extract, the pink pad changes colour to purple-blue. The darker the blue colour, the more glucose is present. Which solution contains most glucose? Record your results.

8 Take the four Albustix strips. Dip one into each tube, then lay the strips on the bench surface beside the tubes they were dipped in.

Albustix is a test for proteins. If protein is present in the extract, the yellow pad changes colour to green. The darker the green colour, the more protein is present. Which solutions contain proteins, and which do not? Record your results.

9 Use the 1 cm³ plastic syringe to add 1 cm³ of iodine solution to each tube. Shake the tubes and allow 2–3 minutes for the contents to settle. This treatment stains starch grains purple-blue. Which solvents have extracted starch? Record your results.

QUESTIONS

1 Draw a table to show how the particles separated in each solvent. (6)

2 Which tube contained
  (a) most oatmeal at the top (1)
  (b) most oatmeal at the bottom (1)
  (c) roughly the same amount of oatmeal at the top and bottom? (1)

3 List the solvents in decreasing order of glucose concentration (strongest first). (4)

4 List the solvents in decreasing order of protein concentration. Which solvents did not extract any protein? (4)

5 List the solvents in decreasing order of starch concentration. (4)

6 How would you test the hypothesis that some of the solvents might affect the response of Clinistix to glucose and Albustix to protein? (4)

Taking it further

1 One manufacturer, 'Silky Cereals', claims that its flour contains the finest grains. Starting with four different makes of flour, two white and two brown, how would you test the manufacturer's claims? Plan and carry out your investigation, writing out your method and results.

2 Repeat this investigation using a pea or bean flour (e.g. soya bean flour). Compare your results. How do you account for any differences?

9 Particle size in cereal products

Investigation: *40–60 mins*

Manufacturers of milled cereal products such as porridge oats, wholemeal flour and white flour need to know the size of particles in their product. If one batch of produce is to be similar to others, the particle size must be carefully controlled. The size of particles can be found by passing the product through a series of sieves with holes of different sizes. Starting with a known quantity, the amount of material retained by each sieve is measured and recorded.

This investigation can be carried out using a sieving outfit consisting of a set of nesting sieves mounted in aluminium frames with an aluminium lid and receiver. Aperture sizes are as follows:

Sieve no.	Aperture (mm)
1	1.7
2	0.85
3	0.25
4	0.18

The material retained by each sieve is weighed, together with any material which is in the receiver after using the sieve with the smallest mesh.

Investigation

Materials

- 50 g porridge oats
- 50 g wholemeal flour
- 50 g white flour
- 100 cm^3 beaker (container for weighing samples)
- top-pan balance
- nest of sieves, with lid and receiver

Method

1 Take sieve no. 1, with the receiver in position underneath. Tip the porridge oats into the sieve. Put on the lid, then gently rock the sieve from side to side until no more material will pass through the mesh (see Fig 1).

2 Weigh the beaker. Tip the contents of sieve no. 1 into the beaker. Measure and record the mass of material exceeding 1.7 mm in diameter.

3 Repeat instructions 1 and 2 with material that has passed through sieve no. 1, using sieves 2, 3 and 4 in turn. Weigh and record the mass of material retained by each sieve, together with any material that has passed through sieve no. 4.

4 Repeat the above instructions using wholemeal flour and white flour. Record all your results.

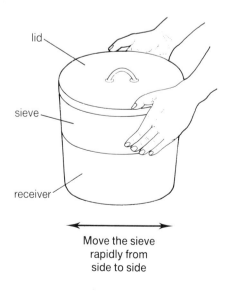

Fig 1

QUESTIONS

1 Record all of your results in the form of a table. (5)

2 Plot your results as a histogram. (8)

3 Which product contained most particles
   (a) larger than 1.7 mm
   (b) between 0.25 and 0.85 mm
   (c) smaller than 0.18 mm? (3)

4 If larger particles pass through the alimentary canal faster than small ones, which product would pass through
   (a) fastest
   (b) slowest? (2)

5 Name two possible sources of error in your results. (2)

Taking it further

1 Compare the size of particles in (a) two batches of porridge oats from the same manufacturer and (b) porridge oats from 3 to 5 different manufacturers. How does the size of particles affect the appearance of the product, and the time taken for cooking?

2 Bran is the coarse, brown covering of wheat grains, normally retained by sieve nos. 1 and 2. It adds roughage to the diet and speeds up the rate at which food passes through the alimentary canal. Try to find out how much bran is present in different types of flour, including wholemeal and white. Express your results as a percentage of dry mass.

10 Physical properties of milk

Investigation: *50–70 mins*

Milk is the raw material from which different dairy products are made. Cow's milk, for example, contains **protein** (casein), **carbohydrate** (lactose) and **fat**. The fat is present in the form of minute white droplets, which float to the surface where they form cream. Although most people like cream, there is some evidence that the saturated fats which it contains can cause heart disease. Producers of dairy products are now separating the fat from the milk to meet consumer demand for healthier foods. The cream is skimmed off to leave a less fattening and less harmful skimmed milk.

In this investigation you will use two methods for measuring the cream content of milk. You will also find out how different amounts of cream affect the rate at which milk cools, an important consideration in some industrial processes such as condensed milk production.

Investigation

Materials

- 70 cm^3 full cream milk
- 70 cm^3 skimmed milk
- 2 test tubes fitted with rubber bungs
- 2 100 cm^3 beakers, graduated in 25 cm^3 units
- 50 cm^3 measuring cylinder
- 10 cm^3 plastic syringe
- filter funnel
- 2 thermometers
- retort stand, boss and clamp
- scalpel
- ruler, graduated in mm
- glass-marking pen
- matchstick
- waterbath maintained at 90°C
- stop-clock, or watch with a second hand

Method

1 Draw a horizontal ink mark 5 cm above the base of each test tube.

2 Pour full cream milk into one of the test tubes up to the ink mark. Label the tube and fit a bung. Shake the tube vigorously up and down for 2–3 minutes until butterfat floats to the surface. Measure and record the depth of the butterfat.

3 Repeat instruction 2 using skimmed milk. Measure and record the depth of butterfat.

4 Set up the retort stand, boss and clamp. Remove the plunger from the syringe. Use the scalpel to shape a small piece of matchstick so that it fits loosely into the nozzle of the syringe.

Investigations in Applied Biology and Biotechnology © 1990 Peter Freeland. Published by Hodder & Stoughton

29

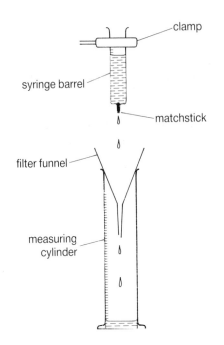

Fig 1

Fill the syringe with water and adjust the position of the stick so that it releases 80–120 drops per minute. Clamp the syringe barrel into position. Place a filter funnel and measuring cylinder beneath the syringe (Fig 1).

5 Pour full cream milk into the syringe barrel to a level above the 10 cm³ mark. Start the stop-clock as soon as the milk level falls to 10 cm³. Measure and record the milk level every 30 seconds until the syringe is empty.

6 Repeat instruction 5 using skimmed milk.

7 Measure 50 cm³ of full cream milk and pour it into one of the beakers. Label the beaker. Pour 50 cm³ of skimmed milk into the other beaker. Label it.

8 Transfer the beakers to the waterbath, maintained at 90°C. Leave them in the waterbath for 10–15 minutes.

9 Put a thermometer into each beaker. Carefully remove the beakers from the waterbath and stand them on the bench. Without stirring or moving the positions of the thermometers, measure and record the temperature of each milk sample at 2-minute intervals, until the temperature of each sample has fallen to 30°C.

QUESTIONS

1 **(a)** What was the depth of butterfat in (i) full cream and (ii) skimmed milk? (2)
(b) Calculate the percentage butterfat in (i) full cream and (ii) skimmed milk. Show how you arrived at your answers. (2)

2 Copy and complete the table. (6)

	Vol. milk (cm³)			
	Full cream		Skimmed	
Time (seconds)	In syringe	Drained out	In syringe	Drained out
0				
30				
60				
90				
120				
150				
180				
⋮				
300				

3 **(a)** Plot graphs to show the flow rates of (i) full cream and (ii) skimmed milk. Use the same axes for both graphs. (5)
(b) What is the relationship between flow rate and cream content? (2)

4 **(a)** Draw a table to show the rate at which (i) full cream and (ii) skimmed milk cooled. (4)
(b) What do you conclude from your results? (4)

Taking it further

1 Design and draw a milk separation plant suitable for the large-scale separation of cream from milk. Your metallic separator has a capacity of 5000 litres, with inlet and outlet pipes controlled by stopcocks.

2 Devise an accurate method for finding out how much cream is present in a milk sample. How could you find out if water had been added to a milk sample?

11 When does custard thicken?

Investigation: 30–40 mins

Precautions

Heat the custard very slowly, often withdrawing the Bunsen burner. Do not allow the mixture to reach boiling point, or it will boil over. Hold hot syringes by the handle and plastic flange at the top of the barrel, otherwise your fingers may be burned.

Custard is usually made by mixing custard powder with milk, then heating the mixture until it thickens. Thickening occurs because the custard powder contains starch in the form of small, hard grains. Heat from the hot milk breaks hydrogen bonds in the starch molecules. As a result they open up, take in water and swell to form a **gelatinous** mass. This process is sometimes called **gelatinisation**.

In this investigation you will try to find out the temperature range over which gelatinisation takes place. **You must take great care and follow the safety procedures carefully.**

Investigation

Materials

- 2 g custard powder
- 2 100 cm^3 beakers
- 5 cm^3 plastic syringe
- thermometer
- tripod and gauze
- Bunsen burner
- stop-clock, or watch with a second hand
- eye protection

Method

1 Pour 100 cm^3 of water into one of the beakers. Tip the custard powder into the water and stir the mixture with the thermometer. Record the temperature.

2 Use the syringe to draw up 5 cm^3 of the mixture. Set your stop-clock or watch. Hold the syringe, nozzle pointing down, over the empty beaker, and gently pull the plunger out of the barrel. Measure and record the time (seconds) it takes for the syringe to drain (Fig 1).

3 Put on your eye protection. Set up the tripod and gauze. Light the Bunsen burner. Adjust the gas tap and burner to produce a low, steady flame. Place the beaker of custard mixture on the gauze, above the Bunsen burner. Stir the mixture and heat it gently until the temperature reaches 25°C. Remove the Bunsen burner.

4 Carefully withdraw 5 cm^3 of the mixture and record the time it takes to drain from the syringe.

5 In the same way, measure and record the time it takes for the mixture to drain from the syringe after heating to 35, 45, 55, 65, 75, 85 and 95°C.

Take care: observe all the safety precautions when handling the hot mixture.

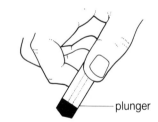

plunger

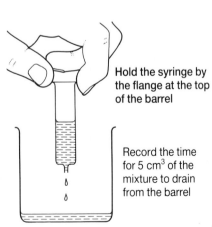

Hold the syringe by the flange at the top of the barrel

Record the time for 5 cm^3 of the mixture to drain from the barrel

Fig 1

QUESTIONS

1 Copy and complete the table. (8)

Temperature (°C)	Time for drainage of syringe (s)
room	
25	
35	
45	
55	
65	
75	
85	
95	

2 Present your results in the form of a graph. (5)

3 Over what range of temperature did gelatinisation take place? (2)

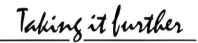

1 Starch from different plants undergoes gelatinisation at different temperatures. Try to find out if cornflour, packet soups and brown flour gelatinise at different temperatures.

2 Another gelatinising agent is agar. Use a 2%$^{w/v}$ solution of agar to find the temperature at which it gelatinises. As the molten agar cools, it hardens and solidifies into a gel. At what temperature does this happen?

3 Find out more about the use of gelatinising agents in the food industry. Write an essay or prepare a talk for your class on your findings.

12 Making and using yogurt

Investigation: 30–40 mins

Precautions

Do not eat any yogurt prepared in the laboratory.

Yogurt is milk which has **coagulated** into soft curd. In the commercial preparation of yogurt, two kinds of bacteria, *Lactobacillus bulgaricus* and *Streptococcus thermophilus* (also known as *Lactococcus thermophilus*), are used to ferment the milk. *Lactobacillus* ferments lactose into lactic acid. This acid then coagulates milk protein, called casein, into a solid curd. *Streptococcus* is used mainly to give the yogurt a creamy texture and improve the flavour.

In this investigation you will make yogurt by culturing bacteria in sterilised milk.

Investigation

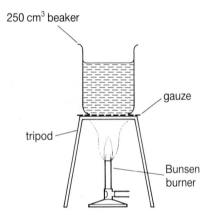

Fig 1

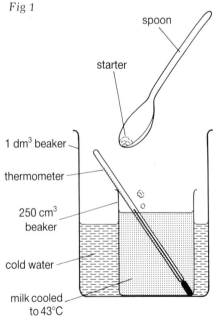

Fig 2

Materials

- 200 cm³ milk in a plastic beaker
- yogurt 'starter' culture or teaspoon of live yogurt
- 250 cm³ beaker
- 1 dm³ beaker
- thermometer
- teaspoon
- cloth (for handling hot beakers)
- Bunsen burner, tripod and gauze
- incubator maintained at 43°C
- access to refrigerator
- Clingfilm
- eye protection

Method

1 Put on your eye protection. Set up the tripod, gauze and Bunsen burner, as shown in Fig 1. Pour about 200 cm³ of tap water into the 250 cm³ beaker. Light the Bunsen burner and heat the water until it boils. Carefully tip away the boiling water.

2 Pour the milk into the 250 cm³ beaker. Heat the milk to 82°C, and keep it at this temperature for 2–3 minutes. Pour about 300 cm³ of cold water into the 1 dm³ beaker. Cool the heated milk to 43°C by placing the 250 cm³ beaker inside the 1 dm³ beaker, as shown in Fig 2. When the temperature reaches 43°C stir in the 'starter' culture. Continue stirring until the culture is mixed in evenly.

3 Cover the 250 cm³ beaker with Clingfilm to keep out micro-organisms from the air. Put the beaker in an incubator at 43°C and allow 3–4 hours for the milk to ferment. (The longer yogurt is left at this temperature, the more acid it will become.)

4 Transfer the yogurt to a refrigerator. Allow 3–6 hours for the yogurt to cool and gel.

QUESTIONS

1 (a) Explain the reason for boiling water in the 250 cm³ beaker before adding the milk. (1)
(b) What might have happened to the yogurt if this step had been left out? (1)

2 Why was the milk heated to 82°C? (1)

3 Suggest two reasons for not heating the milk until it boiled (100°C). (2)

4 Why was the milk cooled to 43°C before the 'starter' culture was added? (1)

5 What was the reason for limiting fermentation to 3–6 hours? (1)

6 A student who used a live yogurt as a 'starter' was disappointed to find that after 3–4 hours in an incubator the milk still had not thickened. Suggest two possible reasons for this. (2)

7 Why should you not eat yogurt prepared in the laboratory? (1)

Taking it further

1 You can eat yogurt prepared at home. Heat some milk to 82°C in a saucepan, cool to 43°C, then stir in the 'starter'. Take a wide-necked Thermos flask and sterilise it by filling it with hot water (85–100°C). Empty the flask and allow it to cool, then half-fill it with the yogurt mixture. Replace the screw top and allow the mixture to ferment for 2–10 hours. Tip out and cool the yogurt as soon as the mixture has thickened, but before it becomes too acid.

What were the advantages of preparing the yogurt in a wide-necked Thermos flask?

2 The yogurt you have made at home can be flavoured by adding chocolate powder or canned fruit and sugar. Alternatively, you can use it to make coleslaw, by mixing it with other ingredients in the following proportions:

yogurt	: 5 parts
shredded white cabbage	: 10 parts
mayonnaise	: 3 parts
grated carrot	: 2 parts
lemon juice	: 1 part
salt to taste	

Devise and prepare other dishes which contain yogurt.

Investigations in Applied Biology and Biotechnology © 1990 Peter Freeland. Published by Hodder & Stoughton

13 Cheese-making

Investigation: Part 1 – 30–50 mins
Part 2 – 10–15 mins
Part 3 – 20–40 mins

Precautions

Do not eat cheese prepared in the laboratory.

Cheese-making is an ancient biotechnology. People think it originated in ancient Egypt, Macedonia and parts of Asia. Milk was allowed to sour naturally or **coagulate** until solid curd had separated from liquid whey. After draining off all the whey, the curd was mixed with salt, pressed into moulds, cooled, dried and left to mature.

Modern cheese-making is a more carefully controlled process. Milk can now be coagulated rapidly by adding **rennet**, an enzyme preparation extracted from the stomachs of calves or from fungi. More often, sterilised milk is treated with a 'starter' culture consisting of different species of lactose-fermenting bacteria. Each species gives a cheese its characteristic flavour, which can be strengthened by letting fungi grow on the maturing curd.

In this investigation, after using rennet and a 'starter' culture to coagulate milk and give a characteristic flavour, you will go through all the stages of cheese-making.

PART 1

Materials

- 1 pint milk
- cheese 'starter' culture
- 10 cm^3 rennet solution
- large freezer bag
- large saucepan
- thermometer
- string
- access to cooker

Method

1 Pour the milk into the freezer bag. Tie the neck of the bag with the string.

2 Half-fill the saucepan with water. Put the freezer bag containing milk into the saucepan. Slowly heat the water in the saucepan until it boils. Turn off the heat and cool until the water temperature has reached 30–35°C.

3 Carefully untie the freezer bag. Add the 'starter' culture and rennet solution to the milk. Re-tie the bag. Shake it to mix the contents.

4 Stand the bag in a warm place for 12–24 hours.

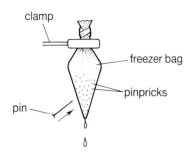
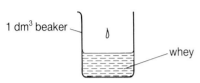

Fig 1

PART 2

Materials

- coagulated milk (in freezer bag)
- 1 dm³ beaker
- retort stand, boss and clamp
- pin

Method

1 Set up the retort stand, boss and clamp. Support the freezer bag containing milk in the clamp. Put the 1 dm³ beaker under the bag.

2 Using the pin, pointing upwards, make 50–300 pinpricks in the bag (Fig 1). Allow 3–12 hours for the curd to drain. Gently squeeze the freezer bag from time to time to assist drainage.

PART 3

Materials

- cheese curd (in freezer bag)
- 2 g salt
- muslin, approximately 15 × 15 cm
- aluminium foil, approximately 15 × 15 cm
- cardboard cylinder, approximately 8 cm long × 6 cm wide
- teaspoon
- access to boiling water

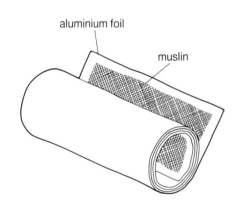

Fig 2

Method

1 Untie the string. Add salt to the curd and mix it in with the spoon.

2 Sterilise the muslin by dipping it into boiling water.

3 Lay the aluminium foil on the bench. Place the muslin on top of it. Roll up the aluminium foil, with the muslin inside, to fit inside the cardboard cylinder (Fig 2).

4 Fold one end of the aluminium foil inwards to form the base of a container for the salted curd. Use the spoon to fill the container with salted curd (Fig 3). Press it firmly into position. Fold the free end of the muslin inwards, to cover the top of the curd.

5 Invert the container and store it for 2–3 weeks at 15–20°C in a pie dish or similar container. After 2–3 weeks, close the open end of the aluminium foil cover. Store it for a further 2–3 weeks and then examine the colour and texture of the cheese.

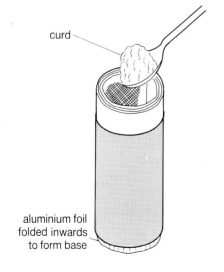

Fig 3

Investigations in Applied Biology and Biotechnology © 1990 Peter Freeland. Published by Hodder & Stoughton

QUESTIONS

1 What is:
  (a) the name of the sugar in milk (1)
  (b) the general name of bacteria capable of fermenting this sugar (1)
  (c) the name of the acid produced after fermentation? (1)

2 What enzyme is present in rennet? (1)

3 Distinguish between:
  (a) 'unripened' and 'ripened' cheeses (2)
  (b) soft and hard cheeses. (2)

4 What are the main aims of 'ripening' a cheese? (2)

1 Explain the reasons for each of the following stages in cheese-making:
sterilising milk; adding a 'starter'; adding rennet; draining the curd; salting; moulding; storing before eating.

2 The hardness of curd, and of the cheese made from it, depends partly on the amount of rennet added to the milk. Design and carry out an investigation to find out how different amounts of rennet affect the hardness of cheese.

3 Find out more about commercial cheese production. Draw a diagram to show the stages in the commercial production of a hard cheese such as Cheddar. How does the manufacture of a soft cheese, such as Lymeswold, differ?

14 Cheese tasting

Investigation: 15–25 mins

Imagine that you work for a cheese company called Cheddartaste which sells Cheddar cheeses to supermarkets. The supermarkets want to stock only those cheeses that are popular with customers so that they sell quickly. You've been asked to carry out a survey of customers' preferences, analyse the results, and to recommend what cheeses to stock.

This is an investigation that can be carried out on a group of five or more pupils. It is best to carry out the tasting at home, not in the school laboratory. After you have collected your results, you can begin to think about the best way of analysing and presenting your findings.

Investigation

Materials

- 5 or more 'cheese-tasters'
- New Zealand Cheddar, labelled A
- farmhouse Cheddar, labelled B
- mature Cheddar, labelled C

Method

1 You have three different Cheddar cheeses. Taste the cheeses and list them in order of preference, with the one you liked best first.

2 Taste a little more of cheese A. From the following list of words, choose the two that best describe this cheese: peppery; salty; mild; creamy; musty; nutty; gritty; sickly; smooth; acid.

Write down your choice.

3 Now select two words from the list that best describe cheeses B and C. Write down your choices.

4 Collect all the responses from the group.

QUESTIONS

1 Draw a table to represent the responses of all the pupils in your group. (4)

2 Devise and use a numerical scoring system to find out which cheese was
  (a) most popular
  (b) least popular
with the tasters. Explain the method you used. Write down any additional tables, data, etc. that you used to get a result. Present your results in the form of a table. (5)

3 Count and record the number of times that each word from the list was used to describe each of the cheeses. Present your results in the form of a table. (4)

4 Suppose you wanted to find out which feature of each cheese people liked (i) best and (ii) least. Given the same list of words, what instructions would you give to the tasters? How would you analyse your results? (4)

5 Criticise the method used and suggest how it might be improved. (3)

Taking it further

1 Flat Fizz, a brewing company, wants to market a non-alcoholic beer. Write down a list of 5–10 words that you think tasters might use to describe this product.

2 An alternative way of getting information about food products is to use a questionnaire, with questions about colour, flavour, etc. Try to design a questionnaire, with not more than ten questions, to find out what people think of a well-known brand of chocolate bar.

Investigations in Applied Biology and Biotechnology © 1990 Peter Freeland. Published by Hodder & Stoughton

15 Acid and sugar in citrus fruits

Investigation: *40–60 mins*

Oranges, lemons, limes, tangerines and grapefruit are citrus fruits. They are all closely related species of small trees which grow in sub-tropical and tropical climates. Citrus fruits contain large amounts of **citric acid**, which gives them their sharp, refreshing taste. As they ripen, the fruits become sweeter, because of the presence of glucose and fructose in the juice.

Shoppers prefer thin-skinned citrus fruits, with plenty of juice. Growers need to monitor the juice content of their fruits and to pick them at the right time, when glucose and fructose concentrations are high.

In this investigation you will measure the amount of juice that can be extracted from an orange and a lemon. You will also find out how much citric acid and **reducing sugar** (glucose and fructose) the fruits contain.

Precautions

Use forceps when handling Clinitest tablets.

Investigation

Materials

- orange
- lemon
- 2 Clinitest tablets
- Clinitest colour chart
- 35 cm^3 0.1 M NaOH solution
- 25 cm^3 1%$^{w/v}$ citric acid solution
- phenolphthalein indicator in a dropping bottle
- 2 flat-bottomed tubes or test tubes in a rack
- 2 250 cm^3 beakers
- 100 cm^3 beaker
- 100 cm^3 measuring cylinder
- 10 cm^3 plastic syringe
- 5 cm^3 plastic syringe
- 1 cm^3 plastic syringe
- scalpel
- forceps
- glass-marking pen
- top-pan balance

Method

1 You have an orange and a lemon. Weigh each fruit and record its mass.

2 Cut the orange across the middle into two halves. Squeeze all of the juice from the two halves into one of the 250 cm^3 beakers. Label the beaker. Tip the juice into the measuring cylinder. Measure and record its volume. Tip the juice back into the 250 cm^3 beaker.

3 In the same way, cut and squeeze the lemon, collecting the juice in the other 250 cm^3 beaker. Label the beaker. Use the measuring cylinder to measure the volume of juice from the lemon. Record the volume.

Investigations in Applied Biology and Biotechnology © 1990 Peter Freeland. Published by Hodder & Stoughton

4 You have a 1%$^{w/v}$ citric acid solution and a 0.1 M NaOH solution. Use the 10 cm^3 syringe to put 10 cm^3 of NaOH solution into the 100 cm^3 beaker. Add a few drops of phenolphthalein indicator until the mixture is pink/red. Fill the 5 cm^3 syringe with citric acid solution and add it, drop by drop, to the NaOH solution, until the mixture becomes colourless. Record the volume of citric acid solution added.

5 Wash out the beaker. Put 10 cm^3 of NaOH solution into the beaker and add phenolphthalein until the mixture is pink/red. Fill the 5 cm^3 syringe with orange juice and add it, drop by drop, to the NaOH solution until the mixture becomes colourless. Record the volume of orange juice added.

6 In the same way, find the volume of lemon juice needed to neutralise 10 cm^3 of NaOH solution. Record your results.

7 Use forceps to put one Clinitest tablet into each of the flat-bottomed tubes or test tubes. Take the 1 cm^3 plastic syringe and draw in 0.7 cm^3 of orange juice. Add this to a Clinitest tablet in one of the tubes. Allow 1–2 minutes for the reaction to take place. Use the colour chart to read off the approximate concentration of glucose in the mixture. Record your results.

8 Use the second Clinitest tablet to find the approximate concentration of glucose in the lemon juice. Record your results.

QUESTIONS

1 What was the mass of
   (a) the orange
   (b) the lemon? (2)

2 What was the volume of juice squeezed from
   (a) the orange
   (b) the lemon? (2)

3 Assuming that 1 cm^3 of juice weighs 1 g, calculate the percentage juice in
   (a) the orange
   (b) the lemon.
Show how you arrived at your answers. (4)

4 What was the volume of citric acid solution required to neutralise 10 cm^3 of NaOH solution? (1)

5 (a) What was the volume of orange juice required to neutralise 10 cm^3 of NaOH solution?
   (b) How much citric acid is present in 100 cm^3 of orange juice? Show how you arrived at your answer. (2)

6 (a) What was the volume of lemon juice required to neutralise 10 cm^3 of NaOH solution? (1)
   (b) How much citric acid is present in 100 cm^3 of lemon juice? Show how you arrived at your answer. (1)
   (c) What assumption have you made? (1)

7 What was the approximate concentration of reducing sugar in
   (a) the orange juice
   (b) the lemon juice? (2)

8 Citric acid and reducing sugars have different effects on the taste buds. Citric acid tastes sour; reducing sugars taste sweet. After looking at all your results, try to explain why lemon juice has a sourer taste than orange juice. (6)

Taking it further

1 Use the same method to find which of the citrus fruits (a) provides most juice/100 g of fruit, (b) contains most citric acid/100 cm^3 of juice and (c) contains most reducing sugar/ 100 cm^3 juice.

Are there differences in the juice, citric acid and glucose contents of oranges (or lemons) from different countries?

2 (a) Find out how much citric acid and reducing sugar there is in cartons of fruit juice. Look at the list of contents on each packet. Has citric acid been added? Does the product contain glucose syrup?

Find out how much it would cost to produce 1 litre (1000 cm^3) of pure orange juice.

(b) Make a list of products stocked by your local supermarket which citric acid has been added to. Try to find out how this citric acid is produced.

Investigations in Applied Biology and Biotechnology © 1990 Peter Freeland. Published by Hodder & Stoughton

16 Making sweeter sugar

Investigation: 50–80 mins

Precautions

Use forceps when handling Clinitest tablets. Do not hold tubes containing Clinitest/sugar solution mixtures as they heat to 100°C.

Investigation

We add sugar, a natural sweetener, to many foods. There are many different sugars. The most widely used is sucrose, which comes from sugar cane and sugar beet. Although sucrose is abundant, it isn't very sweet and so it needs to be added in fairly large amounts. Evidence suggests that eating large amounts of sucrose is harmful to your health. It contributes to obesity, high blood pressure and tooth decay. One aim of biotechnology is to produce other sugars, including glucose and fructose, from sucrose. Fructose has a sweeter taste than sucrose, so it can sweeten foods when added in smaller amounts. One effect of this could be better health.

Sucrose can be converted to a mixture of glucose and fructose by adding the enzyme **invertase**, which is produced by yeast:

$$\text{sucrose} \xrightleftharpoons{\text{invertase}} \text{glucose} + \text{fructose}$$

The resulting mixture of sugars tastes sweeter than sucrose. If you simply add invertase to a sucrose solution you have to remove it from the final products before they can be used. This is quite difficult and could be expensive in an industrial process. Biochemists have therefore developed a method of avoiding this problem by attaching the enzyme to a surface so that it stays in one place and does not mix with the glucose and fructose. This is called **immobilisation**. Another advantage is that the enzyme can be used over and over again.

In this investigation you will immobilise invertase by trapping it in jelly-like beads of sodium alginate.

Materials

- 5 cm³ invertase concentrate
- 30 cm³ 3%$^{w/v}$ sodium alginate solution in a beaker
- 50 cm³ 3%$^{w/v}$ calcium chloride solution in a beaker
- 150 cm³ 2%$^{w/v}$ sucrose solution
- 5 Clinitest tablets
- Clinitest colour chart
- 250 cm³ beaker
- 5 specimen tubes, approximately 8 × 1.5 cm
- 10 cm³ plastic syringe
- retort stand, boss and clamp
- kitchen sieve
- freezer bag
- test tube holder
- forceps
- glass rod
- pin
- stop-clock, or watch with a second hand

Method

1 Add the invertase concentrate to the sodium alginate solution and stir it with the glass rod.

2 Fill the syringe with the mixture. Hold the syringe nozzle over the beaker of calcium chloride solution and slowly depress the plunger until drops of the mixture fall into the beaker. Continue this process until all the enzyme/sodium alginate mixture has been changed into beads. (These beads contain the immobilised enzyme.)

3 Pour the calcium chloride solution, and the beads it contains, through the sieve. Retain the beads, and wash them under a running tap.

4 Transfer the beads to the plastic freezer bag. Pour the sucrose solution into the bag. Tie the neck of the bag and suspend it from the clamp, as shown in Fig 1.

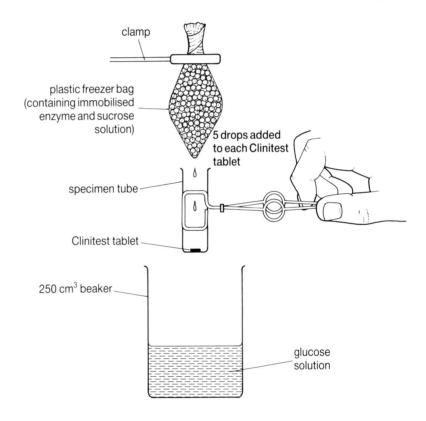

Fig 1

5 Stand the 250 cm³ beaker under the bag. Use the pin to make one or more small holes at the base of the bag to release 20–40 drops per minute. Record the time when the apparatus was set up.

6 Use forceps to put one Clinitest tablet into each of the specimen tubes. Five minutes after setting up the drip, take one of the specimen tubes in the test tube holder. Hold the tube under the bag until five drops have fallen onto the tablet. Remove the tube, rest it on the bench, and observe the colour after the reaction has stopped. Read off and record the approximate concentration of glucose from the colour chart.

7 In the same way, measure and record the approximate amounts of glucose in the mixture after 10, 15, 20 and 25 minutes.

QUESTIONS

1 (a) Draw a table to show changes in glucose concentrations over a period of 25 minutes.
(4)

(b) Plot your results as a graph. (5)

2 What must you do to the glucose solution to extract solid glucose? (2)

3 State two problems you would have to solve before the apparatus could be scaled up into an industrial plant for producing pure glucose. (4)

4 Design and draw a simple production plant for producing glucose from sucrose using apparatus commonly found in a laboratory. (5)

Taking it further

Immobilisation is most useful in reactions catalysed by enzymes that are either expensive or difficult to obtain. Furthermore, the products obtained are relatively free from contamination by enzymes. The following table shows some of the reactions in which immobilised enzymes, or living cells, could be used to obtain useful products.

Enzyme substrate	Enzyme(s)	Product
paper (cellulose)	cellulase + cellubiase	glucose
starch	amylase + maltase	glucose
proteins	pancreatic enzymes	peptides + amino acids

Look up Investigation 24, which decribes the use of immobilised enzymes in sand columns. Set up sand columns, similar to that shown on sheet 24A, to produce one or more of these end products. What reagents could be used to detect the presence of these products?

Investigations in Applied Biology and Biotechnology © 1990 Peter Freeland. Published by Hodder & Stoughton

17 Energy content of foods

Investigation: 40–50 mins

Food supplies the energy required to drive all the chemical reactions that take place in our cells. A normal diet contains three energy-giving nutrients: carbohydrates, fats and proteins. Europeans get about 40 per cent of their energy from carbohyrdrates, 45 per cent from fats and 15 per cent from proteins. The amount of energy contained in foods can be expressed in terms of calories. A **calorie** (with a small 'c') is the amount of heat required to raise 1 cm^3 of water through 1°C, while a **Calorie** (with a large 'C') is the amount of heat required to raise a litre of water through 1°C. Hence, 1 Calorie = 1000 calories. One gram of both carbohydrates and proteins supply 4.1 Calories of energy. A gram of fat, however, supplies 9.3 Calories. An average adult has a daily energy requirement of about 1800 Calories. If energy intake, as food, exceeds energy expenditure, any excess food is converted into fat and stored. This can lead to obesity (see Investigation 36).

The calorific values of foods can be found by burning them beneath a given volume of water and measuring the rise in temperature of the water. In this investigation you will compare the energy content of (a) dry-roasted and oil-roasted peanuts and (b) cream cracker and digestive biscuits.

Investigation

Materials

- ½ dry-roasted peanut, labelled A
- ½ oil-roasted peanut, labelled B
- piece of cream cracker biscuit, labelled C
- piece of digestive biscuit, labelled D
- 4 test tubes
- test tube rack
- 10 cm^3 plastic syringe
- thermometer
- mounted needle
- glass-marking pen
- retort stand, boss and clamp
- Bunsen burner
- eye protection

Method

1 You have the following foods:
   ½ dry-roasted peanut, labelled A
   ½ oil-roasted peanut, labelled B
   piece of cream cracker biscuit, labelled C
   piece of digestive biscuit, labelled D.
Your aim is to burn each of these foods, in turn, beneath a test tube containing water.

2 Label the tubes A, B, C and D. Use the syringe to put 10 cm^3 of water into each tube.

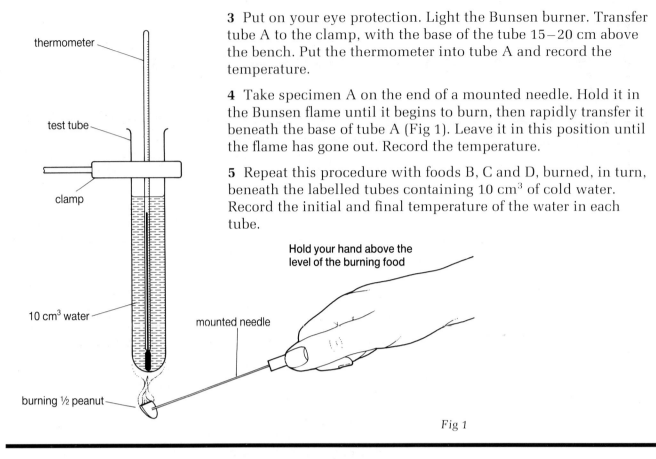

3 Put on your eye protection. Light the Bunsen burner. Transfer tube A to the clamp, with the base of the tube 15–20 cm above the bench. Put the thermometer into tube A and record the temperature.

4 Take specimen A on the end of a mounted needle. Hold it in the Bunsen flame until it begins to burn, then rapidly transfer it beneath the base of tube A (Fig 1). Leave it in this position until the flame has gone out. Record the temperature.

5 Repeat this procedure with foods B, C and D, burned, in turn, beneath the labelled tubes containing 10 cm³ of cold water. Record the initial and final temperature of the water in each tube.

Fig 1

QUESTIONS

1 Present your results in the form of a table. Include one column for the increase in temperature. (5)

2 A calorie is the unit of heat required to raise 1 cm³ of water through 1°C. Calculate the number of calories in each of the foods A, B, C and D. Show how you arrived at your answers. (4)

3 The term 'calorie' is no longer used by scientists, who prefer to measure energy output in joules. It takes 4.2 joules of heat energy to raise the temperature of 1 cm³ of water by 1°C. Calculate the number of joules in each of the foods A, B, C and D. Show how you arrived at your answers. (4)

4 How do you account for any differences between the heat energy released from:
 (a) the two peanuts (2)
 (b) the two biscuits? (2)

5 The method you have used is inaccurate for a number of reasons. List three inaccuracies, and suggest ways in which the design of the investigation could be improved. (3)

Taking it further

The apparatus shown in Fig 1 is unsuitable for foods such as butter, cheese and chocolate. These foods would melt when heated, and run off the needle.

Using familiar laboratory apparatus such as beakers, crucibles, tripods, thermometers, gauzes, Bunsen burner, etc. design a piece of apparatus suitable for finding the calorific value of a piece of chocolate. Draw your apparatus. Write down, in the order they should be carried out, instructions for using the apparatus.

18 'E' is for artificial colouring

Investigation: *45–60 mins*

Some manufacturers add artificial colouring to foods and drinks in order to improve their appearance, making them more attractive to customers. These **artificial colourings**, along with other food additives are coded with an E on food labels. Before any colouring is added to food for humans it is tested extensively for safety by feeding it to animals. Even so, some people are known to be allergic to these compounds. Furthermore, many people object to adding chemical colourings to food because the long-term effects of these coloured pigments on humans are not known.

In this investigation you will use thin-layer chromatography to separate and identify the pigments in blue and pink food colouring. You will then use this information to find out if any of these artificial colourings are present in blackcurrant squash.

Investigation

Materials

- blue food colouring (labelled A)
- pink food colouring (labelled B)
- blackcurrant squash (labelled C)
- 10–15 cm³ chromatographic solvent
- 250 cm³ glass beaker
- 3 strips of silica-coated plastic, each 6.5 × 1.0 cm
- 3 paintbrushes
- ruler, graduated in mm
- pencil

Method

1 You have three strips of silica-coated plastic. Take the pencil and draw a faint horizontal line about 1.0 cm from one end. Press *lightly* on the pencil, taking care not to remove any of the white silica powder.

2 You have three paintbrushes, and the following coloured solutions:

blue food colouring (labelled A)
pink food colouring (labelled B)
blackcurrant squash (labelled C)

Take one of the paintbrushes, dip it into the blue food colouring, and paint a thin horizontal line to cover the pencil mark on one of the strips. Blow gently on the colouring until it is dry. Lightly pencil the letter A below the line.

3 Paint a thin horizontal line of pink food colouring on one of the remaining strips, and blackcurrant squash on the other. Use a clean paintbrush for each one. Dry and label the strips.

4 Pour the solvent into the beaker to a depth of 0.5–0.75 cm. Carefully transfer the painted strips to the beaker. Rest them against the side of the beaker, with the painted line just above the solvent. Space out the strips so that they do not touch.

Investigations in Applied Biology and Biotechnology © 1990 Peter Freeland. Published by Hodder & Stoughton

49

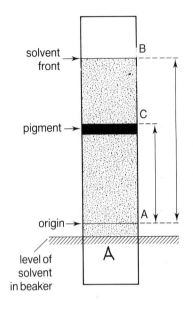

5 Leave the solvent to rise up the sheets for 10–15 minutes, so that the pigments separate from each other. Carefully remove the strips and mark the position of the solvent front (Fig 1) with a pencil. Rest the strips, pigments facing upwards, on the bench to dry.

6 Using a ruler, measure and record the following distances on each strip:

(a) The distance (mm) between the origin and solvent front (distance A–B in Fig 1)
(b) The distance (mm) between the origin and leading (upper) edge of each pigment (distance A–C in Fig 1).

7 From your results calculate the relative flow rates (R_f values) of each pigment.

$$R_f = \frac{\text{distance moved by pigment (A–C)}}{\text{distance moved by solvent (A–B)}}$$

Fig 1

QUESTIONS

1 What were the colours and R_f values of the pigments present in
 (a) solution A (2)
 (b) solution B (1)
 (c) solution C? (1)

2 Which pigment travelled through the silica powder
 (a) fastest
 (b) slowest? (2)

3 After examining all of your chromatograms, what do you conclude about the colouring used in the blackcurrant squash? (2)

4 Why must you not remove any of the silica powder when marking the origin? (1)

5 Why must the solvent front be marked immediately after removing the strips from the solvent? (1)

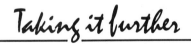

1 *E for Additives* by M Hanssen published by Thorsons Press, is a recent book that describes all of the artificial colourings added to food and drink. Obtain a copy of this book from your school or public library. Look up carmoisine, etythrosine and brilliant blue. Write a short account of the chemical nature of these compounds, and some of the adverse effects they have had on people.

Why are these dyes added to food and drink? Do you think the practice should be continued? Give reasons for your views. What are the alternatives?

2 Supermarkets also sell green, yellow and brown food colourings. What pigments do these colourings contain? How would you find out if artificial colouring had been added to (a) canned cherries and (b) tomato sauce?

19 Natural colouring

Investigation: *40–60 mins*

Precautions

Take care! The chromatography solvent is highly inflammable and toxic.

The possibility that some artificial colourings may be a long-term threat to human health has led to a search for natural alternatives. You might think that any product labelled 'contains natural colouring' or 'no artifial colourings' was additive-free. This, however, is not correct. The term '**natural colouring**' refers to a plant or animal product, as opposed to one derived from coal tar or other sources. For example, strawberry jam with 'natural colouring' contains an orange-red plant pigment which rarely, if ever, comes from the strawberry plant itself. Most probably it has come from an alga, grown commercially in large water tanks. After harvesting, the algal pigments are removed by crushing, and separated by chromatography. Finally, the pigments are dissolved in a solvent and sold to food manufacturers or pharmaceutical companies making vitamin pills and other health products.

In this investigation you will extract and compare the number and colour of pigments from (a) stinging nettle leaves and (b) *Spirulina*, or another alga.

Investigation

Materials

- leaf of stinging nettle
- culture of *Spirulina* (or another alga)
- 10 cm^3 acetone in a stoppered container
- 2 cm^3 chromatography solvent in a stoppered container
- 2 TLC plastic/silica powder strips, each 2 × 8 cm
- 250 cm^3 beaker
- pestle and mortar
- aluminium foil (to cover beaker)
- paintbrush
- forceps
- pencil (lead)
- coloured pencils (orange, red-pink, yellow, yellow-green, blue-green)
- ruler, graduated in mm
- tissue paper

Method

1 Put the stinging nettle leaf into the mortar. Add 2–3 cm^3 of acetone. Use the pestle and mortar to grind the leaf to a slurry.

2 Take one of the silica-coated strips and draw a thin horizontal line about 1 cm from one end.

3 Dip the paintbrush into the slurry and paint a thin line over the pencil mark. (You can add more green slurry as the line dries, but try not to increase the thickness of the line.)

Investigations in Applied Biology and Biotechnology © 1990 Peter Freeland. Published by Hodder & Stoughton

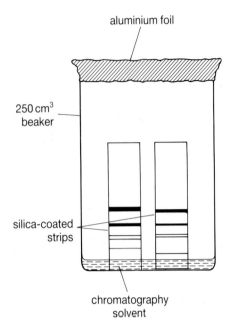

Fig 1

4 Pour the chromatography solvent into the beaker. Carefully transfer one silica-coated strip to the beaker, with the green line just above the solvent. Cap the beaker with aluminium foil (Fig 1).

5 Use the tissue paper to clean the paintbrush and the pestle and mortar thoroughly.

6 Take the culture of *Spirulina*, remove the alga with forceps, and repeat instructions 1–4. Transfer the second silica-coated strip to the beaker, positioning it so that it does not touch the first strip. Replace the aluminium cap. Allow 15–20 minutes for the pigments to separate.

7 Remove the strips from the beaker. Mark, with the lead pencil, the position of the solvent front on each strip.

QUESTIONS

1 Draw outline shapes of the two silica-coated sheets. Label them (i) stinging nettle and (ii) *Spirulina*. Use coloured pencils to show the colour and arrangement of pigments on each strip. (8)

2 (a) In stinging nettle, how many (i) yellow and (ii) green pigments were present?
(b) How many (i) orange-red and (ii) green pigments were present in *Spirulina*?
(c) Name a colouring, sometimes added to ice-cream or sweets, that could not be obtained from either plant. (10)

3 The relative flow rate of a pigment, or its R_f value, is given by the following equation:

$$R_f = \frac{\text{distance pencil line–pigment front (cm)}}{\text{distance pencil line–solvent front (cm)}}$$

Use this information to find out if the green pigments extracted from the two plants are identical. Explain your method and give your results. (5)

4 R_f values are affected by the chemical nature of the solvent. List two other factors that might affect R_f values. (2)

Taking it further

1 Use the same solvent and silica-coated sheets to find out how many different pigments can be extracted from orange peel. Are these pigments the same as those found in (a) lemon peel, (b) carrot root and (c) leaves?

2 Try to extract pigments from flower petals with (a) water and (b) ethanol. Attempt to separate the water-soluble pigments by paper chromatography, using an HCl:methanol (1:9) solvent mixture. Separate the ethanol-soluble pigments on silica gel sheets, with cyclohexane/ethyl acetate as solvent.

20 Watching chloroplasts at work

Investigation: *30–50 mins,*
March–September

When biologists want to study a process such as **photosynthesis**, it is sometimes more convenient to use specially treated cell extracts rather than whole plants or cells. Photosynthesis is the process by which plants manufacture their food. It is summarised in the following equation:

$$6CO_2 + 6H_2O \xrightarrow[\text{chlorophyll}]{\text{light}} C_6H_{12}O_6 + 6O_2$$

carbon   water   glucose   oxygen
dioxide

Oxygen is a by-product. The rate at which this gas is released from leaves shows how fast they are photosynthesising. Measuring oxygen production from leaves, however, can be difficult. The main problem is that the leaves must be put into water in order to trap the bubbles of oxygen. Unfortunately, this treatment may cause oxygen production to fall, as it can interfere with photosynthesis.

This investigation shows one way in which a new technology, that of trapping chloroplasts inside jelly-like beads of sodium alginate, can be used to investigate the effects of light intensity on the rate of photosynthesis.

Investigation

Materials

- 50 stinging nettle leaves, freshly picked
- 50 cm^3 10%$^{w/v}$ sucrose solution
- 10 cm^3 3%$^{w/v}$ sodium alginate solution in a stoppered test tube
- 100 cm^3 beaker containing 50 cm^3 3%$^{w/v}$ calcium chloride solution
- 100 cm^3 beaker
- 5 cm^3 plastic syringe
- 100 cm^3 measuring cylinder
- petri dish
- glass rod
- spoon
- ruler, graduated in cm
- bench lamp fitted with 100 W bulb
- kitchen mixer
- stop-clock, or watch with a second hand

Method

1 Put the stinging nettle leaves and sucrose solution into the kitchen mixer. Grind the leaves until they are reduced to a slurry. Take about 10 cm^3 of this slurry and put it into the empty beaker. Add the sodium alginate solution. Mix it with the glass rod.

Investigations in Applied Biology and Biotechnology © 1990 Peter Freeland. Published by Hodder & Stoughton   53

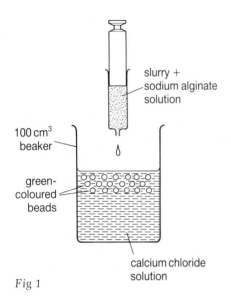

Fig 1

2 Draw up some of the mixture into the syringe. Hold the syringe vertically, with the nozzle about 10 cm above the beaker of calcium chloride solution (Fig 1). Gently push down on the plunger to release a steady stream of drops into the calcium chloride solution. Green-coloured beads should form. Stop after you have made about 70 beads. Use the spoon to remove the beads. Put them into the petri dish.

3 Fill the measuring cylinder with water to the 100 cm³ mark. Put 10 beads into the measuring cylinder. Switch on the bench lamp. Place it on the bench, with the bulb 10 cm from the measuring cylinder, pointing directly at the beads. Read the time. Record the time taken for 5 and 10 beads to rise to the surface.

4 Record the time taken for 5 and 10 beads to rise when the bulb is 15, 20, 25, 30 and 35 cm from the measuring cylinder.

QUESTIONS

1 Record your results in the form of a table. (6)

2 Plot your results as a graph. (4)

3 (a) Your graph shows that the rate of photosynthesis falls with decreasing light intensity. From the figures in your table, draw a second (i) table (ii) graph to show how the rate of photosynthesis increases with increasing light intensity. (Light intensity is calculated from the equation $\frac{1}{d^2} \times 1000$, where d = distance (cm) of lamp from the measuring cylinder.) (10)

(b) Explain how you got the figures for your second table. (3)

4 Name two sources of error in your investigation. (2)

Taking it further

1 Design and carry out an investigation using chloroplast-containing beads of sodium alginate to find out how the rate of photosynthesis, at constant light intensity, is affected by temperature.

2 Beads of sodium alginate will rise in water when they contain materials that are releasing a gas. These materials could be enzymes, plant or animal cells, yeasts or bacteria. You can monitor various reactions in the materials by watching the beads rise. Think of at least two ways in which beads of sodium alginate could be used in a simple investigation. Write down the method you would use for each investigation, and attempt to carry it out.

21 Bread-making

Investigation: Part 1 – 70–90 mins
Part 2 – 30–40 mins

Precautions

Do not eat any bread prepared in the laboratory.

The main ingredient in bread is flour, which is made by grinding wheat, barley or rye grains. As early as 4000 BC the Egyptians discovered that bread could be made lighter, tastier and more digestable by adding **yeast** to the dough. We know now that the yeast produces enzymes which catalyse two reactions in the dough:

(a) fermentation of sugars, mostly maltose, to form traces of alcohol and bubbles of carbon dioxide. These bubbles become trapped in the thick, sticky dough, making it increase in volume or 'rise';
(b) breakdown of tough, stringy proteins (glutens) into smaller, more easily digested molecules.

Modern bread-making involves three stages called **kneading**, **fermentation** and **baking**. Firstly, all the ingredients are mixed thoroughly, or kneaded, to form a dough. The dough is then left to ferment at a temperature of 35–40°C until its original volume has doubled. Finally, after shaping the dough into loaves or rolls, it is baked in an oven at 220–250°C.

The following instructions are for making a basic wholemeal bread.

PART 1

Materials

- 250 g wholemeal flour
- 8 g soya flour
- 8 g activated yeast
- 2 g malt extract (or molasses)
- 2 g sodium chloride
- 150 cm^3 warm water
- 5 cm^3 sunflower oil
- mixing bowl
- 250 cm^3 beaker
- wooden spoon
- teaspoon
- Bunsen burner, tripod and gauze
- thermometer
- Clingfilm
- waterbath maintained at 35–40°C
- eye protection

Method

1 Put on your eye protection. Set up the Bunsen burner, tripod and gauze. Pour 150 cm^3 of water into the beaker. Light the Bunsen burner and heat the water to 40°C.

2 Remove the beaker. Add the yeast, malt extract and soya flour to the warm water. Stir with the glass rod until the ingredients are thoroughly mixed together.

3 Put the wholemeal flour and salt into the mixing bowl. Add the yeast/malt extract/soya flour mixture, together with the sunflower oil. Stir with the wooden spoon until a dough forms.

4 Using your hands, knead the dough. Fold the edges towards the centre, trapping some air in each fold (Fig 1). Continue kneading for 5–10 minutes.

5 Cover the bowl with Clingfilm. Transfer the bowl to a waterbath at 35–40°C. Allow 1 hour for the dough to rise.

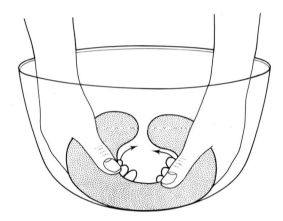

Fig 1

PART 2

Materials

- bread dough in mixing bowl
- baking tray
- oven maintained at 220–250°C

Method

1 Remove the bowl from the waterbath and take off the Clingfilm. Knead the dough for a further 2–4 minutes.

2 Divide the dough into two equal portions. Put each portion on a baking tray, and shape them into loaves or rolls.

3 Transfer the tray to the oven. Bake at 220–250°C for 20–30 minutes, until a crust forms and the bread sounds hollow when tapped.

QUESTIONS

1 Name the micro-organism used in bread-making. (1)

2 List the other ingredients used in bread-making. Which of these ingredients is the source of enzymes? (3)

3 What is the difference between 'leavened' and 'unleavened' bread? (2)

4 What are the main reasons for 'leavening' dough? (3)

5 What happens to the alcohol formed during the fermentation of dough? (1)

Taking it further

1 You have made wholemeal bread. In addition to the wholemeal flour, several other ingredients were mixed together to make the dough. What were the reasons for adding each of the following: yeast; malt extract; soya flour; sunflower oil; salt?

2 Find out more about commercial bread production. Make a drawing to show the stages in the production of white sliced bread.

3 Many different types of bread are now available in shops. Try to find out how one or more of the following types of bread are made: soft crust; granary; soda; rye; savoury; pizza; croissant; brioche; pumpernickel; pitta; muffin; crumpet.

22 Wine-making

Investigation: Part 1 – 30–40 mins
Part 2 – 60–90 mins
Part 3 – 30–40 mins
Part 4 – 20–30 mins
March–May

Precautions

Do not drink wine prepared in the laboratory.

Wine-making is an ancient biotechnology, first used more than 10 000 years ago. Wine can be made from many different starting materials, including fruits, root vegetables, herbs and flowers. These materials serve mainly as a source of flavour. After the flavour has been extracted by boiling with water, yeast and sugar are added. The fermentation of sugar by enzymes from yeast produces alcohol (ethanol) and carbon dioxide:

$$\underset{\text{glucose}}{C_6H_{12}O_6} \xrightarrow{\text{enzymes}} \underset{\text{alcohol}}{C_2H_5OH} + \underset{\text{carbon dioxide}}{CO_2}$$

When fermentation is complete, the wine is cleared, bottled and stored until it is mature and ready for drinking.

This investigation includes the four main stages in wine-making:

flavour extraction → fermentation → clearing → bottling

By following the instructions you can make dandelion flower wine, adding extra flavour from the rinds of lemons and oranges.

Investigation

PART 1

Materials

- 650 g dandelion heads
- 1%$^{w/v}$ sodium hypochlorite solution (for sterilising utensils)
- large saucepan (3–5 litre capacity)
- plastic bucket (covered)
- access to cooker

Method

1 Pick 650 g of dandelion heads (traditionally gathered on St. George's Day, April 23rd).

2 Sterilise the plastic bucket by filling it with sodium hypochlorite solution. Leave the solution to stand for 5–10 minutes. Rinse out the bucket with tap water. Replace the lid.

3 Put the dandelion heads into a large saucepan. Add 2 litres of water. Heat the mixture until it boils. Carefully pour the boiled mixture into the plastic bucket. Replace the lid. Allow the dandelion heads to soak in the water for 2–3 days.

PART 2

Materials

- soaked dandelion heads
- 1 kg sucrose
- rinds of 2 lemons and 1 orange

- activated wine yeast
- thermometer
- wooden spoon
- large saucepan (3–5 litre capacity)
- plastic bucket (covered)
- access to cooker

Method

1 Return the dandelion head/water mixture to the saucepan and boil for 1 hour. Pour the boiled mixture into the bucket.

2 Put the sucrose and rinds into the saucepan. Add $2\frac{1}{2}$ litres of water and heat until the mixture boils. Carefully pour the boiled mixture into the bucket.

3 Cool the boiled mixture to 20–25°C. Add the activated yeast and stir the mixture with the wooden spoon. Replace the lid of the bucket. Let it stand for 2–3 days.

PART 3

Materials

- 2 g ammonium sulphate
- 2 g ammonium phosphate
- $1\%^{w/v}$ sodium hypochlorite solution (for sterilising utensils)
- gallon jar
- fermentation lock
- muslin bag
- jug

Method

1 Sterilise the gallon jar and fermentation lock in sodium hypochlorite solution. Rinse in water.

2 Strain the contents of the bucket through the muslin bag into the jug or a similar container. Pour the strained extract into the gallon jar. Add the ammonium sulphate and ammonium phosphate. Fit the fermentation lock and stand the jar in a warm place (20–25°C) for 1–4 weeks, until bubbles of gas have stopped passing through the fermentation lock.

PART 4

Materials

- raw wine (in gallon jar)
- 5 g pectinase powder
- $1\%^{w/v}$ sodium hypochlorite solution (for sterilising utensils)
- 4 wine bottles
- 4 screw-caps (or corks)
- filter funnel
- jug
- kitchen towel

Method

1 If the wine is hazy, add 5 g of pectinase powder to the contents of the gallon jar. Stir, and allow the mixture to stand at 21–27°C for 4 hours.

2 Sterilise the wine bottles and screw-caps (or corks) with sodium hypochlorite solution and wash them in tap water. Filter the wine through a kitchen towel into the jug.

3 Pour the wine into bottles, fit screw-caps (or corks) and store them in a cool, dark place for 6 months.

QUESTIONS

1 (a) Name the micro-organism used in wine-making. (1)
(b) Is the same organism used to make beer? Explain your answer. (2)

2 Commercial wine producers bubble sulphur dioxide (SO$_2$) through raw grape juice, or 'must', before it is inoculated with yeast. Explain the reason for this. (2)

3 Wine produced by the natural fermentation of sugars may contain anything from a few per cent up to 16 per cent alcohol. What prevents more alcohol from being produced? (2)

4 Explain each of the following terms used to describe wines:
(a) dry (1)
(b) fortified (1)
(c) spirit. (1)

Taking it further

1 Explain the reason for each of the following stages in making dandelion wine: washing apparatus in hypochlorite solution; using freshly-picked flower heads; boiling the heads; boiling the sugar solution; adding yeast at or below 25°C; adding pectinase powder; filtering the wine; storing before drinking.

2 Wines can be made from a large number of starting materials. Some of the most widely used are listed below.

Flowers: elderflower; gorse; pansy; primrose; rose.
Fruits: apple; apricot; banana; bilberry; blackberry; cherry; currant; fig; gooseberry; grape; mulberry; orange; peach; plum; prune; raspberry; rhubarb; sloe; strawberry.
Root vegetables: carrot; beetroot; parsnip; pea pod; potato.
Herbs: mint; parsley; thyme.

Buy or borrow a book on wine-making. Using proper wine-making equipment, and sterile techniques, try to broaden your experience of wine-making.

3 How are beer and lager made? Draw a diagram of the stages involved in commercial brewing. What are the problems caused by scaling up beer production?

60   Investigations in Applied Biology and Biotechnology © 1990 Peter Freeland. Published by Hodder & Stoughton

23 Malting barley

Preparation: *20–30 mins*
Investigation: *20–30 mins, over 6–8 days*

Beer is made from barley. At harvest, barley grains contain mostly starch. There is very little sugar for yeast to change into alcohol during fermentation. Brewers have known for many years that by allowing barley grains to germinate, most of the starch is changed into a sugar. We now know that this sugar is **maltose**, and the enzyme which catalyses the transformation is called **amylase**.

$$\text{starch} \xrightarrow{\text{amylase}} \text{maltose}$$

The process of germinating barley grains is called **malting** and the product, which can be dried and roasted, is called malted barley. After crushing and adding hot water, the extract from the malted grains, or **malt**, is fermented by yeast to produce beer.

In this investigation you will measure the amount of reducing sugar, mostly maltose, produced by germinating barley grains. You will also compare the production of amylase in two regions of a barley grain, known as the embryo and endosperm halves (Fig 1).

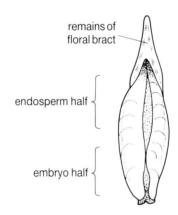

Fig 1

Precautions

Use forceps when handling Clinitest tablets.

Preparation

Materials

- 20 g barley grains
- starch/agar/iodine plate
- Clinitest tablets
- Clinitest colour chart
- plastic lunch box
- 100 cm³ beaker, graduated
- flat-bottomed tube, approximately 10 × 2 cm
- 1 cm³ plastic syringe
- pestle and mortar
- scalpel
- forceps
- cotton wool
- sand

Method

1 Line the lunch box with cotton wool. Add water until the cotton wool is moist. Sow the barley grains on the moist cotton wool.

2 Take two barley grains and identify the endosperm and embryo ends (see Fig 1). Cut the two barley grains in half across the middle. Open the starch/agar/iodine plate and put the half-grains, cut surface downwards, on the plate, as shown in Fig 2.

3 Put 40 grains into the mortar, add a pinch of sand, and grind the grains until they are broken into small pieces. (This may take some time.) Scrape the broken grains into the beaker. Add 5 cm³ water. Stir the mixture with your syringe.

4 Use forceps to put a Clinitest tablet into the flat-bottomed tube. Draw up 0.7 cm³ of the solution from the beaker into the 1 cm³ syringe. Add this to the Clinitest tablet. Wait for a reaction to take place, and then read off the approximate concentration of reducing sugar from the manufacturer's colour chart. Record your results.

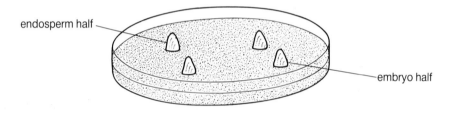

Fig 2

Investigation

Materials

- plastic lunch box with germinating barley grains
- starch/agar/iodine plate with half-grains of barley
- Clinitest tablets
- Clinitest colour chart
- 100 cm³ beaker, graduated
- flat-bottomed tube, approximately 10 × 2 cm
- 1 cm³ plastic syringe
- pestle and mortar
- forceps
- ruler, graduated in mm
- sand

Method

1 Each day, over a period of 6–8 days, use the method described above to measure and record the approximate concentration of reducing sugar in 40 crushed grains, after adding water to 50 cm³. Record all your results.

2 Again, each day for 6–8 days measure and record the diameter of the clear, starch-free zones that develop around each half-grain of barley. (If necessary, flood the surface of the plate with more iodine solution to make these areas clearer.)

QUESTIONS

1 Copy and complete the table. (The area of the starch-free zone is calculated from the formula πr^2, where r = radius and π = 3.14.) (6)

Day	Approx. concentration reducing sugars (%)	Mean diameter of starch-free zone (cm)		Area of starch-free zone (cm²)	
		Endosperm half	Embryo half	Endosperm half	Embryo half
1					
2					
3					
4					
5					
6					
7					
8					
9					
10					

2 (a) From the data in the fourth column of the table (area of starch-free zone), calculate daily amylase production in the two half-grains. Present your results in the form of a table. (4)
(b) What do you conclude from your results? (2)

3 Draw a graph to show how amylase production in the embryo half-grain and maltose production in the whole grain are related. (Plot 'days' along the x-axis. Use the left-hand side of the y-axis for maltose concentrations and the right-hand side for areas cleared of starch. Label your graphs 'maltose' and 'amylase'.) (8)

4 (a) Discuss the relationship between amylase and maltose production. (4)
(b) Make a drawing of a germinating barley grain when the malting process is complete. (4)

5 Name two possible sources of error in your results. (2)

Investigations in Applied Biology and Biotechnology © 1990 Peter Freeland. Published by Hodder & Stoughton

Taking it further

If you have a light meter, a light source and the right chemicals, you can monitor more closely the disappearance of starch and the appearance of maltose in the malted grains.

Set up the light source at a fixed distance above the light meter. Prepare $20^{w/v}$ solutions of (a) starch and (b) maltose. Add 1 cm^3 of iodine solution to 5 cm^3 of starch solution, and add a Clinitest tablet to 5 cm^3 of maltose solution. Use the light meter to measure the absorbance of each solution. Dilute each solution to give concentrations of 1.5, 1.0, 0.5 and 0.25%$^{w/v}$. Again, measure and record the absorbance of each solution after adding iodine solution and a Clinitest tablet. Plot standard curves for starch and maltose.

Sow barley grains on moist cotton wool. Each day over a period of 5–8 days, remove 40 grains, crush them, add water to 50 cm^3, and measure the amount of (a) starch and (b) maltose in the extracts. Plot graphs to show changes in starch and maltose concentrations.

24 Alcohol production

Preparation: 20–30 mins
Investigation: 30–40 mins

The fermentation of glucose by yeast produces **alcohol (ethanol)**. This process is summarised in the following equation, which also shows the approximate yields of alcohol and carbon dioxide from 180 g glucose:

$$\text{glucose} \rightarrow \text{ethanol} + \text{carbon dioxide}$$
$$C_6H_{12}O_6 \quad C_2H_5OH \quad 2CO_2$$
$$(180 \text{ g}) \quad (92 \text{ g}) \quad (44.8 \text{ litres})$$

Alcohol has many industrial uses. For example, it can be used as a solvent for organic compounds, or as fuel for combustion and rocket engines. When natural supplies of oil and gas begin to run out, industry may turn to alcohol instead. Industrial alcohol, however, needs to be pure and free from yeast cells and water.

In this investigation you will make a small-scale plant for producing industrial alcohol. By trapping yeast cells in a sodium alginate gel you can stop them from contaminating the end product. You can then find the amount of alcohol in the end product by using a hydrometer to measure the relative density.

Preparation

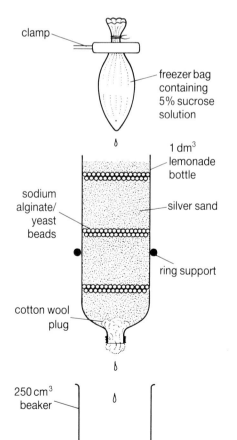

Fig 1

Materials

- 10 g dried yeast
- 100 cm³ 3%$^{w/v}$ sodium alginate solution in a beaker
- 100 cm³ 3%$^{w/v}$ calcium chloride solution in a beaker
- 1 dm³ plastic lemonade bottle (or similar fizzy-drink bottle)
- 10 cm³ plastic syringe
- glass rod
- scissors
- silver sand
- cotton wool

Method

1 Add the dried yeast to the sodium alginate solution, and stir it with the glass rod until the yeast is mixed in evenly. Draw this mixture up into the syringe. Empty it, drop by drop, into the calcium chloride solution to form beads. Repeat this until all the sodium alginate/yeast mixture is converted into beads. Wash the beads and keep them in the beaker until you need them.

2 Use sharp scissors to cut away the base of the lemonade bottle. Plug the neck of the bottle with cotton wool. Turn the bottle upside-down and fill it with alternating layers of silver sand and sodium alginate/yeast beads (Fig 1).

Investigation

Materials

- sodium alginate/yeast bead sand column
- 200 cm^3 5%$^{w/v}$ glucose solution in a freezer bag
- 100 cm^3 beaker
- 100 cm^3 measuring cylinder
- wine hydrometer
- 2 retort stands and bosses
- ring support
- clamp
- pin

Method

1 Set up one of the retort stands with a boss and ring support. Position the inverted bottle as shown in Fig 1 (Sheet 65), with the beaker below the bottle neck. Set up the second retort stand with a boss and clamp, supporting the glucose solution in the freezer bag above the bottle. Use the pin to make a single hole in the bag, to deliver a slow drip of glucose solution through the apparatus. Collect the product in the beaker.

2 Pour the product into the measuring cylinder. Use the wine hydrometer to measure the relative density of the product. Record your result.

QUESTIONS

1 From the equation given in the introduction, calculate
  (a) the theoretical yield of ethanol (g) from 100 cm^3 of a 5%$^{w/v}$ glucose solution (2)
  (b) the theoretical yield of CO_2 from the same volume of 5% glucose solution.
Show how you arrived at your answers. (2)

2 **(a)** In practice, actual yields of ethanol and carbon dioxide are less than the equation predicts. Suggest a reason for this. (2)
  (b) Why do you think home-brewers are advised against fermenting sugars in thin, glass bottles such as beer bottles? (2)

3 The relative density of a 5%$^{w/v}$ glucose solution is approximately 1.015. That of an ethanol/water mixture always less than 1.000.

  (a) What was the relative density of the product from your apparatus? (1)
  (b) What does this tell you about the product? (2)
  (c) What further treatment may be necessary before you obtain an ethanol/water mixture? (1)
  (d) (i) How could you obtain pure ethanol from an ethanol/water mixture? (1)
    (ii) Explain the theoretical basis of the method you choose. (3)

4 **(a)** How could you improve the efficiency of the apparatus? (2)
  (b) List two advantages of using immobilised yeast cells. (2)

Taking it further

1 Fig 2 shows an alternative way of producing industrial alcohol on a small scale. In this apparatus yeast cells are separated from the product by Visking tubing, which is permeable to water and alcohol. Set up the apparatus and use it to investigate changes in the relative density of the liquid outside the Visking tubing.

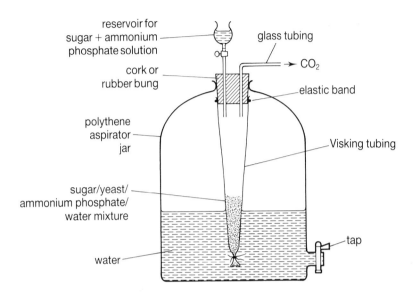

Fig 2

Explain how you would use this apparatus to produce single cell protein (SCP) (see Investigation 26).

2 Vinegar is produced by the aerobic fermentation of an ethanol (10–13%$^{w/v}$)/water mixture by several species of bacteria belonging to the genus *Acetobacter*. You can use the following procedure to set up a small-scale vinegar plant.
Pour a little red or white wine into a flat dish and leave it exposed to the air for 3–4 days, until it smells of vinegar (acetic acid). Mix the soured wine with sodium alginate (3%$^{w/v}$) solution and pipette the slurry, drop by drop, into calcium chloride (3%$^{w/v}$) solution to form beads. Set up an apparatus similar to that shown in Fig 2, replacing the sand with a mixture of polystyrene beads, small pebbles and wood shavings. Drip wine, or an ethanol/water mixture, through the apparatus very slowly. Collect the product.

25 Making a fermenter

Preparation: 40–90 mins

Micro-organisms such as yeasts and bacteria ferment sugars to produce organic acids, alcohols and other compounds. A fermenter reaction vessel is a container in which industrial fermentations can be monitored, controlled and made to operate continuously. Commercially-produced fermentation vessels, or **fermenters**, are expensive. An inexpensive model, suitable for demonstrating the basic principles, can be made from a 2 dm^3 plastic bottle such as a lemonade bottle. When complete, the fermenter has the following parts:

(a) reaction chamber, where fermentation takes place
(b) input point, for adding more substrate (e.g. sugar solution), acid, alkali, bacteria or yeast
(c) sampling tap, for extracting the products of fermentation
(d) probe insertion point, for a thermometer or pH meter
(e) gas outlet.

By following these instructions you can make a simple fermenter, then use it to study the fermentation reactions carried out by bacteria or yeasts.

Precautions

Wear eye protection. Heat/burn the plastic bottle in a fume cupboard. Avoid skin contact with Plastic Padding.

Preparation

Materials

- 2 dm^3 plastic bottle, e.g. lemonade bottle
- 40 cm^3 plastic syringe
- hypodermic needle
- glass tubing, 15–20 cm × 35 mm diameter
- PVC translucent tubing, 3–5 mm diameter and 1.0–1.5 cm diameter
- rubber tubing, to fit over glass tubing
- rubber bung, to fit neck of bottle
- rubber bung, to fit probe insertion point
- Bunsen burner
- bradawl (wooden handled)
- Plastic Padding
- clear nail varnish
- paintbrush
- screw clip
- scalpel
- retort stand, boss and ring
- eye protection
- access to fume cupboard

Method

1 Put on your eye protection. Set up the Bunsen burner in a fume cupboard. Light the burner and turn on the fume cupboard extractor fan. Heat the metal part of the bradawl until it is red hot. Burn a small circular hole in the bottle about 5 cm from the base, and a larger oval hole about 15 cm from the base, preferably on the opposite side (Fig 1).

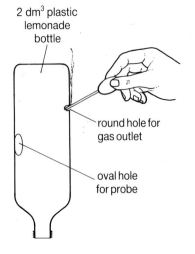

Fig 1

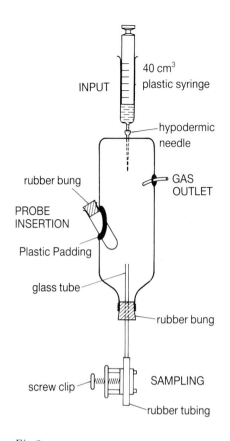

Fig 2

2 Cut a 2 cm length of PVC tubing from each of the samples. Fit the small-diameter tubing into the circular hole, and the large-diameter tubing into the oval one, angled as shown in Fig 2. If necessary, re-shape the holes to make a better fit.

3 Mix Plastic Padding according to the manufacturer's instructions. **Avoid skin contact with the hardener.** Use the paintbrush or scalpel to apply Plastic Padding to the bottle, sealing in the two pieces of PVC tubing. Allow 15–20 minutes for the Plastic Padding to dry and harden. Paint over and around each join with clear nail varnish. Allow a further 15–20 minutes for the nail varnish to dry.

4 Carefully insert the hypodermic needle through the bottom of the bottle. Fit the glass tubing through the rubber bung, then attach a short length of rubber tubing, with a screw clip, to the lower end of the glass tubing (see Fig 2). Complete the fermenter by fitting the small rubber bung at the probe insertion point and the plastic syringe at the input point.

5 Sterilise the completed fermenter, inside and outside, by immersing it in a bath of 1 per cent sodium hypochlorite solution (1% domestic bleach). After immersion for 10–15 minutes, use 2–3 changes of distilled water to rinse out the inside of the fermenter.

6 Set up the retort stand, boss and ring. Support the fermenter in the ring, with the screw clip resting on the bench.

QUESTIONS

1 After sterilising your fermenter with sodium hypochlorite solution, why must it be washed out with water before it can be used? (1)

2 The fermenter you have made is suitable for micro-organisms respiring anaerobically. Name two other features that would have to be fitted if the fermenter was to be used for micro-organisms respiring aerobically. (2)

3 Industrial fermenters are usually made from stainless steel. Suggest three reasons for this. (3)

4 An industrial fermenter may be surrounded by a jacket which superheated water (above 100°C) can be pumped through. What are the advantages of this? (2)

5 Fermentation of glucose by the bacterium *Clostridium acetobutylicum* produces a mixture of solvents and gases (60% butanol, 30% acetone, 7% ethanol, 3% isopropanol, carbon dioxide and hydrogen). How would you separate
   (a) acetone from the other solvents
   (b) hydrogen from carbon dioxide? (2)

Taking it further

Use your fermenter to investigate one of the following reactions:
(a) fermentation of glucose by *Saccharomyces cerevisiae*
(b) fermentation of sucrose by *Saccharomyces cerevisiae*
(c) fermentation of lactose by *Lactobacillus acidophilus*
(d) fermentation of lactose by *Lactobacillus plantarum*.
At timed intervals withdraw 10 cm³ samples from the fermenter, and titrate against 0.1 M sodium hydroxide solution, using methyl orange as indicator. Plot a graph to show how acidity changes with time. How does adding glucose affect the rate of lactose fermentation by *Lactobacillus plantarum*?

26 Cell counts

Preparation: 20–30 mins
Investigation: 30–40 mins, over 6–8 days

In many industrial processes, unicellular organisms are grown as a source of food for domestic animals or humans. One example is **single cell protein (SCP)**. Another is yeast extract, used as a flavouring and source of B vitamins. Growing micro-organisms for food has two advantages:
(a) these organisms have a rapid growth rate, so they produce protein faster than farm animals or crop plants;
(b) bacteria and yeasts can be grown on a variety of substrates, including waste products from the food, agricultural and oil industries. This makes microbial protein relatively cheap.
During a batch culture of a micro-organism, a manufacturer may want to find out how fast a population is growing. This can be done by counting the number of cells in a small volume of culture using a microscope, but this is slow and tedious. A faster method of counting is to use a light meter and to measure the absorbance of a given mass of organisms in a culture.

The aim of this investigation with yeast is to produce a standard curve for the absorbance of different masses of dried yeast, suspended in water. You can then use this standard curve to find the growth rates of two cultures of yeast, each supplied with different nutrients.

Preparation

Materials

- 2 g dried yeast
- 1 cm^3 5%$^{w/v}$ ammonium phosphate solution
- 50 cm^3 5%$^{w/v}$ sucrose solution in a beaker
- 2 beakers, each containing 100 cm^3 water
- 6 petri dishes
- 5 cm^3 plastic syringe
- light meter, lamp fitted with 100 W bulb
- retort stand, boss and clamp
- glass rod
- dropping pipette
- glass-marking pen
- metre rule

Method

1 Set up the lamp, supported by a retort stand, boss and clamp, about 20 cm above the bench (see Fig 1). Switch on the lamp.

2 Switch on the light meter. Put the light-sensitive disc under the lamp. Move the lamp up or down to give a reading of about 100 units on the light meter.

3 Tip the dried yeast into one of the water beakers. Stir the mixture with the glass rod until the yeast cells are evenly mixed.

4 Take a petri dish. Use the syringe to put 5 cm^3 of yeast suspension and 15 cm^3 of water into the dish. Replace the lid of the dish, and label the dish along the side. Use the light meter to measure the absorbance of the mixture (Fig 1). Record results.

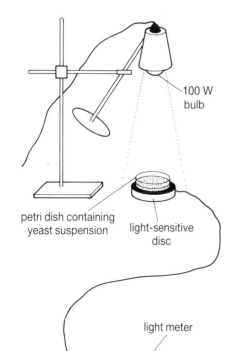

Fig 1

5 Set up petri dishes containing the following mixtures:
   (a) 10 cm³ yeast suspension and 10 cm³ water
   (b) 15 cm³ yeast suspension and 5 cm³ water
   (c) 20 cm³ yeast suspension.
Label each dish along the side. Measure and record the absorbance of each mixture. Record all your results.

6 Set up two petri dishes, each containing 5 cm³ of yeast suspension and 15 cm³ of water. Add 2–3 drops of ammonium phosphate solution to one of the dishes. Label each dish along the side. Leave the dishes on the bench, preferably in a warm place.

Materials

- petri dishes containing yeast suspensions
- light meter
- lamp fitted with 100 W bulb
- retort stand, boss and clamp
- metre rule

Method

1 Set up the lamp and light meter as before, with the lamp switched on and supported about 20 cm above the bench.

2 At two-day intervals over a period of 6 or 8 days, measure and record the absorbance of each yeast suspension.

QUESTIONS

1 Draw a table to show the dry masses of yeast used in the investigation and the absorbance of the suspensions they were contained in. (4)

2 Plot your results as a graph. (6)

3 Make a table to show any increases in the absorbance of the yeast suspension, over a period of 6–8 days, in the petri dishes
   (a) containing ammonium sulphate
   (b) without ammonium sulphate. (4)

4 Plot your results as a graph. Comment on your results. (6)

5 Use your graphs to find the dry mass of the yeast in each petri dish on day 4 of the investigation. State the dry mass of yeast in each dish, and explain how you arrived at your answers. (5)

Taking it further

1 Using the same technique, design experiments to find answers to some of the following questions:
   (a) How does temperature affect the growth rate of yeast?
   (b) What is the effect of pH?
   (c) What elements or compounds are essential for growth?
   (d) What is the pattern of population growth when yeast is grown in a medium containing a fixed amont of nutrients?

2 A different type of light meter has a horizontal beam of light which is passed through a suspension of micro-organisms in a test tube. Compare this type of light meter with the one used in this investigation. Which type is most suitable for measuring the growth rate of micro-organisms? Give reasons for your answer.

27 Which antibiotic?

Preparation: 15–20 mins
Investigation: 30–40 mins

Antibiotics are protective chemicals produced by fungi and bacteria. In the wild these antibiotics are used to kill off rival fungi and bacteria. They allow the organism that produces them to colonise a food source without competition from its rivals. Humans have made use of antibiotics to kill bacteria and fungi that cause disease. Many different kinds of antibiotics are now available to treat bacterial and fungal infections of the blood, lungs, kidneys and other organs. Doctors find the most suitable antibiotic for each patient by growing the disease-causing bacteria or fungi on agar plates, in the presence of different antibiotics. The antibiotic that kills the most organisms on the plate is usually the one chosen to treat the patient.

In this investigation you are going to find out which antibiotics are effective against a common bacterium, *Bacillus subtilis*. You will be given a paper ring called a Mastring, with six different coloured projecting discs, each of which has been dipped into a different antibiotic. Your Mastring is coded as follows:

Colour of disc	Letter code	Antibiotic
green	C	chloramphenicol (25 µg)
red	E	erythromycin (5 µg)
lilac	No	novobiocin (5 µg)
pink	PG	penicillin G (1 unit)
white	S	streptomycin (10 µg)
purple-brown	T	tetracycline (25 µg)

Carry out the instructions to compare the effectiveness of the six different antibiotics.

Preparation

Materials

- petri dish containing nutrient agar
- Mastring-S
- culture of *Bacillus subtilis*
- L-shaped glass spreader
- forceps
- glass-marking pen
- clear adhesive tape
- Bunsen burner
- incubator maintained at 25°C

Method

1 Light the Bunsen burner. Flame the bent end of the glass spreader to destroy any micro-organisms that may be present.

2 Cool the spreader by gently waving it in the air. Carefully dip the spreader into the culture of *B. subtilis*. Hold the petri dish in your other hand. Carefully lift the lid with your thumb and middle finger. As soon as the agar is exposed, spread the culture of *B. subtilis* thinly and evenly over the surface (Fig 1).

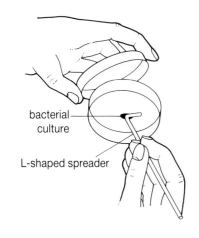

Fig 1

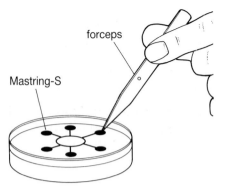

Use forceps to lay the Mastring on the agar. Place it centrally, and press down the coloured paper discs with the tip of the forceps (Fig 2).

3 Seal the dish with clear adhesive tape. Write your name on the bottom of the dish.

4 Turn the plate upside-down and transfer it to an incubator at 25°C. Leave it for 24 hours.

Fig 2

Investigation

Materials

- petri dish containing a Mastring and lawn of *B. subtilis*
- ruler, graduated in mm

Method

1 Examine the petri dish, but do not open it. Use the ruler to measure the diameter of clear (translucent) zones surrounding the paper discs. Record your results.

QUESTIONS

1 Copy and complete the following table. (Use the formula πr^2 to calculate the area of the clear zones, where r = radius and π = 3.14.) (6)

Antibiotic	Diameter of clear zone (cm)	Area of clear zone (cm^2)
chloramphenicol		
erythromycin		
novobiocin		
penicillin G		
streptomycin		
tetracycline		

2 List the antibiotics in order of their activity against *B. subtilis*, with the most active first. (6)

3 The activity of an antibiotic is directly proportional to the area of gel it clears of bacteria. How much more active is chloramphenicol than
  (a) novobiocin
  (b) tetracycline?
Show how you arrived at your answers. (4)

4 Suppose *B. subtilis* caused a disease in humans, and a patient was allergic to chloramphenicol. Which antibiotic would you use, and why? (2)

5 When you prepared the agar plates, what were the reasons for
  (a) sealing the dish with adhesive tape (1)
  (b) incubating the dish upside-down? (1)

Taking it further

1 Spread a plate of nutrient agar with yeast cells. How effective are the antibiotics in controlling the growth of this fungus?

2 Use a different Mastring, impregnated with different antibiotics, to find out how effective they are against *B. subtilis*.

3 Try to find out how antibiotic-resistant bacteria originate, and why they are often so difficult to control.

Investigations in Applied Biology and Biotechnology © 1990 Peter Freeland. Published by Hodder & Stoughton

28 Daily urine output

Investigation: 48 hours

The kidneys are reddish-brown, bean-shaped organs, positioned one on each side of the vertebral column, just above the waist. They receive about a litre (1000 cm^3) of blood per minute from the renal arteries. At the kidneys some of this blood is filtered, its volume is regulated, and waste products, such as **urea**, are removed. Over 24 hours the kidneys of an adult man filter approximately 180 litres of fluid, eliminating about a litre as urine. This yellow fluid, containing urea and other waste products, is stored in the bladder until it is released.

If the kidneys are working normally, water input and output are balanced. If not, fluid is often retained, causing tissues to swell. Measuring urine output is therefore a standard medical test, used by doctors to diagnose kidney disease (see Fig 1). The aim of this investigation is to measure and record your urine output over a period of 48 hours, and then to analyse the results.

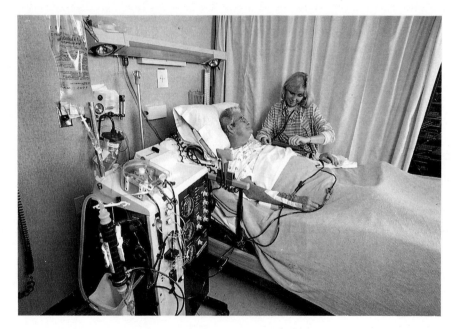

Fig 1 *a patient with kidney failure undergoing dialysis*

Investigation

Materials

- 250 cm^3 measuring cylinder

Method

1 Record the times you urinate over a period of 48 hours. Measure and record the volume of urine produced in each sample.

QUESTIONS

If you have your own results, attempt questions 1–5. If not, use the following results that were obtained by a fifteen-year-old boy.

Day 1		Day 2	
Time	Volume (cm^3)	Time	Volume (cm^3)
07:00	220	02:15	140
09:00	80	09:30	220
11:30	120	12:30	150
14:00	180	15:00	110
17:30	200	17:30	140
22:00	210	19:20	50
		23:00	230

1 Plot your results as a graph (or use the figures given above). (4)

2 Calculate:
(a) total urine output on each day (2)
(b) mean (average) urine output per day (1)
(c) mean urine output per minute. (1)
Show how you arrived at your answers.

3 Over the two-day experimental period, what was the total urine output during the following times:
(a) 06:00–18:00 (1)
(b) 18:00–06:00? (1)

4 From your results, formulate a hypothesis about urine production over a 24-hour period. (1)

5 How would you test your hypothesis? (Include some reference to the size of your sample, their fluid and/or food intake, and the times at which they would be required to urinate.) (4)

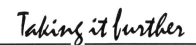

1 Substances which increase urination are called diuretics; those which reduce it are called anti-diuretics. Coffee is a diuretic. Design and carry out an investigation to find out the effect of coffee-drinking on urine output.

Hot chocolate, Ovaltine and Horlicks are popular bedtime drinks. What is the effect of these drinks on urine output? Is it best to take these drinks immediately before going to bed, or earlier in the evening? Design and carry out an investigation to find the answer.

2 The common foxglove (*Digitalis purpurea*) is the source of a drug used to stimulate the heart. Explain, with reasons, the effects you would expect this drug to have on urine output.

29 Drinking and driving

Investigation: 5–10 mins

Alcohol (ethanol) is a sedative drug, with both short-term and long-term effects. Consumed in moderate amounts it removes inhibitions, making people more relaxed and confident. It can sometimes make them more aggressive. Larger amounts cause impaired judgement, dizziness and vomiting. In the long term excessive drinking may cause liver damage, kidney stones and mental disorders.

Today in Britain alcohol addiction, called **alcoholism**, kills twenty times the number of people killed by heroin. It is estimated that one in six people in the UK will become alcoholics. Every year thousands of people are killed or injured in road accidents involving drunk drivers. For example, in 1986 more than 1500 people were killed or seriously injured in alcohol-related accidents over the Christmas–New Year period. The total cost of emergency services dealing with those accidents was £108 million. Caring for the injured in hospitals cost even more. This is why there are such severe penalties for drinking and driving (see Fig 1).

The law states that a motorist can be fined and banned from driving if there is more than 80 milligrams of alcohol in 100 cm^3 of blood. In this investigation, using drinking glasses similar to those shown in Fig 2, you will calculate the volumes of whisky, wine and beer that an average-sized person can drink without exceeding the legal limit. In fact, each glass holds roughly **one unit of alcohol**. If alcohol consumption is to have no long-term effects on health, doctors recommend that a man should not exceed 21 units per week. A woman should not exceed 14 units.

Fig 1 *a motorist being breathalysed by police*

Investigation

Whisky glass	Wine glass	Beer glass
about 35% alcohol	about 12% alcohol	about 4% alcohol

Fig 2

Materials

- whisky glass
- wine glass
- beer glass
- 100 cm^3 measuring cylinder

Method

You have three glasses similar to those shown in Fig 2. Fill each to the brim with water. Using the measuring cylinder, measure and record the total volume of water held by each glass.

QUESTIONS

1 Write down the total volume of each glass. If each container was filled to the brim with the alcoholic drink it is intended for (Fig 2), what volume of alcohol would be present in each glass? Show how you arrived at your answers. (4)

2 The law states that any motorist with more than 80 mg alcohol/100 cm^3 blood can be fined and banned from driving. A 70 kg man has 5 litres (5000 cm^3) of blood in his body. The volume of 80 mg of alcohol = 0.01 cm^3. From these figures, calculate the maximum number of glasses of whisky, wine and beer the man can drink before exceeding the legal limit. Show how you arrived at your answers. (4)

3 In question 2, what assumption is made about alcohol absorption? (1)

4 In general, women have a smaller volume of blood than men. What is the significance of this in relation to drinking and driving? (1)

Taking it further

1 A 70 kg man (blood volume = 5000 cm^3) drinks 5 pints of beer in the Red Lion pub. (1 pint = 473 cm^3.) Should he drive home? What could be the maximum concentration of alcohol in his blood?

2 Supermarkets stock many different types of alcoholic drink. The amount of alcohol in each drink is given on the label as % alcohol/volume. Make a list of alcoholic drinks in increasing order of alcohol content. Which one contains the most alcohol?

Investigations in Applied Biology and Biotechnology © 1990 Peter Freeland. Published by Hodder & Stoughton

30 Variation in a human population

Investigation: 60–90 mins

Every person is different from every other person. Sometimes the differences are because one person has developed a particular skill (e.g. swimming) whilst others have not. Most differences, however, are due to genetic factors, passed on from our parents (**inherited**), such as eye colour (Fig 1). Genes work in pairs. Where only one pair of genes is responsible for a characteristic there are usually very few possible differences. For example, people have either blue eyes or brown eyes. There are no intermediates. This is called **discontinuous variation**. When more than one pair of genes is responsible for a characteristic, such as height, there is a wide range of possible differences. For example, if everyone in your class stands in a line with the shortest at one end and the tallest at the other, then you will see many different heights. This is called **continuous variation**.

This investigation looks at some inherited characteristics. It is useful for doctors to know how people relate to the average height or weight of the population. Furthermore, a doctor may want to know if a person has inherited a gene which could affect his or her health.

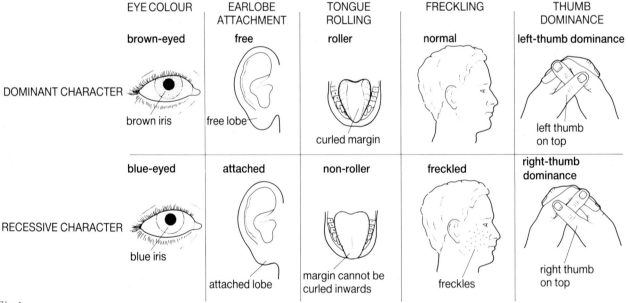

Fig 1

Investigation

Materials

- vertical scale, graduated in cm, for measuring height
- metric scales, for measuring mass

Method

1 Look at Fig 1, which shows some inherited characteristics controlled by single pairs of genes. Find out how many pupils in your class show (a) the dominant and (b) the recessive characteristics of each gene. Record your results.

2 Measure and record (a) the height and (b) the body mass of each pupil in the class. (Shoes should be removed before measuring height; shoes and jackets should be removed before measuring mass.) Collect data for the height and mass of each pupil.

QUESTIONS

1 (a) Copy and complete the table. (5)

Character	No. pupils showing each characteristic		
	Dominant	Recessive	Total
eye colour			
earlobe attachment			
tongue-rolling			
freckling			
thumb dominance			

(b) Make a general statement about the frequency of dominant and recessive forms of these genes. (2)
(c) Plot histograms to show variation in (i) eye colour and (ii) thumb dominance. (4)
(d) What percentage of pupils had (i) blue eyes and (ii) left-thumb dominance? Show how you arrived at your answers. (2)
(e) If you are recessive for any of these characteristics, draw a diagram to show how you inherited the genes from your parents. (2)
(f) If you are dominant for any of these characteristics, show how you could have inherited the genes from your parents. (5)

2 (a) List the heights of pupils in ascending order, starting with the shortest. (2)
(b) What is the mean (average) height of pupils in your class? Show how you arrived at your answer. (2)

(c) Select 4–6 different categories of height (e.g. 120–129 cm, 130–139 cm) to cover the heights of all the pupils in your class. Count and record the number of pupils in each category. Plot a histogram of your results. (6)

3 (a) List the masses of pupils in ascending order. (2)
(b) What is the mean (average) mass of pupils in your class? Show how you arrived at your answer. (2)
(c) Select 4–6 different categories of mass (e.g. 50–59 kg, 60–69 kg) to cover the masses of all the pupils in your class. Count and record the number of pupils in each category. Plot a histogram of your results. (6)

Taking it further

1 Select one of the characteristics controlled by a single pair of genes (Fig 1). Look at the variation in this characteristic within your family, among your brothers/sisters, parents and grandparents. Try to draw a diagram showing how the gene has been inherited over three generations.

2 In a field of buttercups you come across a plant which has flower stalks 35 cm long and six petals per flower. Describe in detail how you could find out if each of these characteristics was controlled by one or several different pairs of genes.

Investigations in Applied Biology and Biotechnology © 1990 Peter Freeland. Published by Hodder & Stoughton

31 Fingerprint analysis

Investigation: *40–50 mins*

The tips of your fingers and thumbs are covered by a pattern of ridges and grooves. The ridges give roughness to your skin, providing grip when things are picked up. They also form your **fingerprints**. Everyone's fingerprints are different, both in the pattern of the ridges and in the width of the grooves. This means that fingerprints can be used to identify people, not only criminals, but also amnesiacs who have forgotten their names, or the victims of fires and accidents.

Fig 1 shows that there are three types of fingerprint patterns, arches, loops and whorls. Each of these types has a number of sub-groups. In this investigation you will find out the approximate frequency of each type of print in (a) girls, (b) boys and (c) the population as a whole.

ARCHES

Plain — parallel ridges gently arch outwards along the longitudinal axis

Tented — there is a central longitudinal ridge with other ridges grouped around it in a triangle-like pattern

LOOPS

Radial / **Ulnar** — there is a semicircular elliptical opening to one side; radial loops open towards the thumb side, ulnar loops towards the little finger

WHORLS

central ridges coil to form ellipses or circles. There may or may not be a central region without ridges (island)

Fig 1

Investigation

Materials

- 5 girls and 5 boys (subjects)
- ink pad
- fingerprint ink
- pen or pencil
- white paper

Method

1 Moisten the ink pad with fingerprint ink. Ask each subject to write their name at the top of the sheet of white paper, then fold the paper transversely into two equal halves.

2 Starting with the right thumb, press firmly on the ink pad, then make a thumb print on the top left-hand side of the paper. Make a print of the right index finger to the right of the thumb print. Follow this with prints of the middle, ring and little fingers. Label these prints 'right hand'.

3 In the same way, make prints of the thumb and fingers of the left hand on the bottom half of the paper. Label these prints 'left hand'.

4 Use Fig 1 to identify the pattern on each finger and thumb. Ask the subjects to write the pattern type above each fingerprint (e.g. tented arch, whorl, etc.).

5 Ask the subjects to add, at the bottom of the paper, 'girl' or 'boy'.

QUESTIONS

1 (a) Write down the print patterns on the fingers and thumbs (digits) of (i) your left and (ii) your right hands. Use abbreviations (e.g. PA = plain arch) for each pattern type. Your thumb is digit no. 1. (4)

(b) How many subjects in your group had the same print pattern on all of their digits? (1)

2 Copy and complete the table. (5)

Fingerprint pattern	No. of each type		Percentage in whole population
	Girls	Boys	
plain arch			
tented arch			
ulnar loop			
radial loop			
whorl			

3 Which fingerprint patterns were:
(a) most common in (i) boys and (ii) girls (2)
(b) least common in (i) girls and (ii) boys? (2)

4 What was:
(a) the percentage of ulnar loops found in (i) girls, (ii) boys and (iii) the population as a whole (3)
(b) the percentage of whorls found in girls? (1)
Show how you arrived at your answers.

5 Which fingerprint patterns were more common in girls than boys? (2)

6 Suggest one or more possible uses of fingerprint analysis. (2)

7 Criticise the design of your investigation and suggest how it could be improved. (3)

Taking it further

Precautions

- -
Remember that silver nitrate is corrosive and can cause skin burns. Wear plastic gloves when handling this reagent.
- -

1 Record your own fingerprints and those of your brothers, sisters, parents and grandparents. Try to construct a diagram, showing how pattern types might be inherited. Do you think print patterns are inherited according to Mendel's laws? Give reasons for your answer.

2 Make four clear, firm fingerprints on a glass slide. Which of the following treatments is best for showing details of a print?
(a) dusting on graphite powder
(b) dusting on cornflour
(c) pouring over a solution of Sudan III or Sudan black dye (0.1%$^{w/v}$ in ethanol)
(d) pouring over silver nitrate solution (0.2%$^{w/v}$).

Investigations in Applied Biology and Biotechnology © 1990 Peter Freeland. Published by Hodder & Stoughton

32 How exercise affects your heart

Investigation: *30–50 mins*

Precautions

Do not attempt this investigation if you have major heart, circulatory or respiratory problems.

Investigation

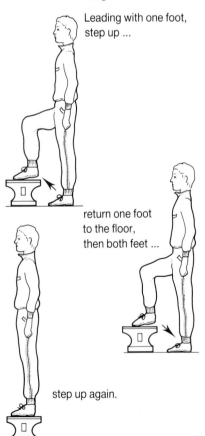

Fig 1

Your heart is a pump, driving blood through your circulatory system. Blood flows from the heart through thick-walled, elastic vessels called **arteries**, which transport blood rapidly to organs and tissues. Once inside a tissue, the arteries divide into smaller **arterioles**, then into extremely small blood vessels called **capillaries**. In the capillaries some fluid flows out to supply the cells with water and nutrients. Oxygen diffuses from the blood to the tissues, while carbon dioxide diffuses in the opposite direction. The blood flows from the tissues in **venules**, and then enters large, thin-walled vessels called **veins** before returning to the heart.

The speed at which blood travels round your body is affected by the size of your blood vessels. If your arteries and veins are partly blocked by fatty deposits of a substance called **cholesterol**, then the blood flow will be slower. To try to overcome this problem, your heart will beat faster and your blood pressure will increase.

In this investigation you will compare the effect of exercise on the heart rates of pupils in your class.

Materials

- gymnasium bench or stool, 25–35 cm in height
- metronome
- stop-clock, or watch with a second hand

Method

The exercise consists of stepping on and off the bench, 30 times a minute, for 3 minutes.

1 Set the metronome at 30 beats a minute. (Alternatively, set up a pendulum made from a weight at the end of a piece of string. Make it swing backwards and forwards at about 30 beats a minute.)

2 Measure and record your pulse rate in your wrist at 2 minutes, and again at 1 minute, before you begin to exercise.

3 Step on and off the bench, 30 times a minute, in time with the metronome. In each movement lead with one foot, place both feet on the bench, return one foot to the floor, then both feet (Fig 1). Continue with the exercise for 3 minutes.

4 Record your pulse rate immediately after exercising and at intervals of 1 minute until your pulse rate has returned to the rate recorded before you began to exercise.

5 Write your pulse rate after exercising, and the time (minutes) taken by your heart to return to its 'resting' rate, on the blackboard so that this information can be used by everyone in the class.

QUESTIONS

1 Present your results in the form of a table. (4)

2 Plot your results as a graph. (6)

3 (a) What happened to your pulse rate immediately before you began to exercise? (1)
  (b) Can you explain this? (1)

4 After exercising, what were
  (a) the highest
  (b) the lowest
pulse rates recorded by members of your class? (2)

5 What was
  (a) the longest
  (b) the shortest
recovery period required by people in your class for their pulse rates to return to normal after exercising? (2)

6 Comment on the significance of the differences you have recorded for questions 4 and 5. (2)

7 Do you think the 'step test' provides useful information about the condition of the heart and circulatory system in young people? Give a reason for your answer. (2)

1 Suppliers of biological equipment now offer inexpensive instruments for measuring pulse rate and blood pressure. Use these instruments to find out how these two factors vary at different times of the day. What is the effect of exercise on blood pressure?

2 If you have a moderately high pulse rate after exercising (e.g. 130–150 beats/min) try to find out if your heart rate can be reduced by (a) dieting and (b) exercising over a period of 2–4 weeks. Which has the biggest effect?

33 Heart and breathing rates during exercise

Investigation: 40–60 mins

You use **voluntary muscles** in all forms of physical activity. When your muscles are active, and you exercise for a long period without a break, you use up more oxygen. This additional oxygen is delivered to the muscles in two ways:
(a) your heart beats faster, and so pumps blood to the tissues at a faster rate;
(b) your rate of breathing (**ventilation**) increases, so that your lungs are filled and emptied more often.
In healthy people there is a general relationship between heart and ventilation rates (see Fig 1). As more oxygen is needed, both heart and ventilation rates are increased automatically.

The aim of this investigation is to find the effect of moderate, sustained exercise, such as stepping on and off a bench, on your heart and ventilation rates. After collecting both sets of results you will think about whether there is a link between them.

Fig 1 athletes running in a long-distance race. Is there a direct relationship between their heart and ventilation rates?

Investigation

Materials

- gymnasium bench
- metronome
- stop-clock, or watch with a second hand
- graph paper

Method

1 Set the metronome at 30 beats per minute. Step on and off the bench, in time with the metronome, for 1 minute. Find the pulse in your wrist or neck, and record your heart rate over a period of 30 seconds. Double this figure to give your heart rate per minute.

2 Concentrate on your breathing, especially the point at which you start to breathe in without consciously changing your breathing rate. Record the number of times you breathe in over a period of 30 seconds. Double this figure to give your ventilation rate per minute. Record your results.

Precautions

Do not attempt this investigation if you have circulatory or respiratory problems.

3 Step on and off the bench at the same rate for further 1 minute periods until you have exercised for a total of 10 minutes. At the end of each 1 minute period, measure and record your heart rate and ventilation rate.

QUESTIONS

1 Draw tables to show how each of the following were affected by exercise:
   (a) ventilation rate (4)
   (b) heart rate. (4)

2 (a) Plot a graph of ventilation rate along the bottom (x-axis) against heart rate up the side (y-axis). (5)

   (b) What is the shape of your graph? (1)
   (c) What conclusions can you draw from the shape of your graph? (2)

3 Suggest two ways in which you could collect more accurate results. (4)

Taking it further

Design and carry out investigations to find the answers to one or more of the following questions:

(a) Does the heart rate vary at different times of the day when people are (i) resting and (ii) exercising?

(b) Is the heart rate affected by (i) sex and (ii) age?

(c) Is the relationship between heart and ventilation rates similar in girls and boys?

(d) What do heart rates and ventilation rates during exercise tell us about people's fitness?

(e) What is the mean (average) ventilation rate of (i) girls and (ii) boys after they have run a 100 m race?

Investigations in Applied Biology and Biotechnology © 1990 Peter Freeland. Published by Hodder & Stoughton

34 Lung capacity

Investigation: 40–60 mins

The lungs are like pumps, drawing air in and out 15–30 times per minute. If you run or take any form of vigorous exercise, the lungs work faster, pumping more air in and out at each breath. The air that is drawn in supplies oxygen, which combines with the red blood pigment **haemoglobin** in the lung's capillaries. The heart then pumps this **oxygenated blood** to the tissues, where it oxidises glucose to produce energy that keeps the muscles moving. Clearly, how fast you can run is controlled by the amount of air that you can take into your lungs. This volume, measured in cm^3, is called the **vital capacity**. Fig 1 shows how vital capacity increases with age between 9 and 17 years. Fig 2 shows a hand (pocket) **spirometer**, used to measure vital capacity. After fitting the mouthpiece you inhale fully, pinch your nostrils until they are closed, and then breathe out through your mouth. The spirometer records the volume of air breathed out. This gives an approximate measure of your vital capacity.

Vital capacity tends to be above average in athletes and other people who take plenty of exercise. On the other hand, it may be below average in those who take little exercise or who smoke. This investigation looks at the condition of the respiratory system in ten people.

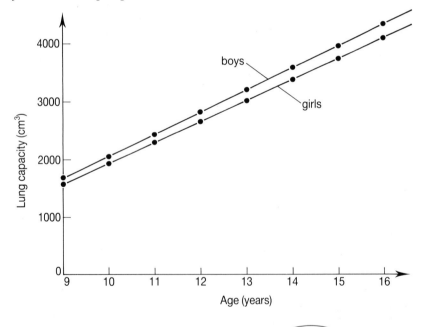

Fig 1 vital capacity of the lungs: average values for 9–17 year olds

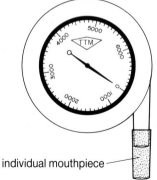

Fig 2

Investigation

Materials

- 10 people (subjects)
- hand (pocket) spirometer
- 10 individual mouthpieces
- metric scales, for measuring body mass

Method

1 Ask the subjects to remove their shoes and jackets or coats. Weigh each subject. Record their names and body masses.

2 Ask each subject to take the spirometer, fit an individual mouthpiece, breathe in, hold their nose, and blow into the spirometer. Record the vital capacity of each subject against their name and body mass.

QUESTIONS

1 (a) Copy and complete columns 1, 2 and 3 of the table. (6)

Subject no. and name	Vital capacity (cm³)	Body mass (kg)	Vital capacity (cm³)/ body mass (kg)
1			
2			
3			
4			
5			
6			
7			
8			
9			
10			

(b) In general, the larger a person, the greater the vital capacity of their lungs. From the data in columns 2 and 3 of the table, devise a simple formula for relating lung capacity to body mass. Use your formula to obtain data for the fourth column of the table. Add this data to the table. (5)

(c) A biologist has suggested using the following formula for the relationship between lung volume and body mass:

$$\frac{\text{vital capacity (cm}^3)}{\text{body mass (kg)} \times 60}$$

(i) Name one advantage of using this formula. (2)

(ii) What is the point of measuring lung volume:body mass ratio? (2)

2 (a) Plot a graph of body mass along the bottom (x-axis) against vital capacity up the side (y-axis) for the ten subjects in your study. (6)

(b) Comment on the relationship. (4)

Investigations in Applied Biology and Biotechnology © 1990 Peter Freeland. Published by Hodder & Stoughton

Taking it further

1 Take monthly readings of your vital capacity over a period of 3–6 months. What is the average monthly increase in your vital capacity? How could you find out if (a) breathing exercises, (b) jogging and (c) colds have a beneficial or harmful effect on your vital capacity?

2 Emphysema is a disease that reduces lung expansion. It may result from prolonged lung irritation, caused by smoking, or by breathing polluted air. Try to find out how the disease is caused and how it reduces the vital capacity of the lungs.

3 Attempt to design apparatus for measuring chest expansion using materials operated either (a) manually or (b) electronically.

4 Compare vital capacity in boys and girls of the same age. Try to explain any differences that you find. Use the following equations to find out if individuals in your group have a better-than-average lung capacity : body mass ratio:

For girls:

$$\frac{\text{vital capacity (cm}^3)}{\text{body mass (kg)} \times 55} = 1.0$$

For boys:

$$\frac{\text{vital capacity (cm}^3)}{\text{body mass (kg)} \times 60} = 1.0$$

Explain the significance of figures that are (a) above and (b) below 1.0.

35 Measuring grip strength

Investigation: 10–15 mins

The strength of human muscles can be measured by using fairly simple, inexpensive apparatus. Fig 1 shows a hand **dynamometer** for measuring the strength of the muscles in the wrist and hand, used to grip objects and to squeeze them. During the time you spend at secondary school (e.g. from 11 to 17) the strength of these muscles usually increases, regardless of how much you use them. (See the table below.) There may be further muscle development in people who play games, swim, or operate tools or machinery.

The aim of this investigation is to analyse the data from each pupil measuring the grip strength of their right and left hands. The table below shows average readings for boys aged 11–17. Those for girls are approximately 5 per cent lower.

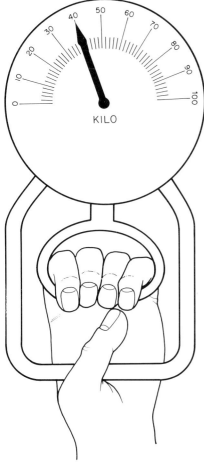

Age (years)	Grip strength (kg) Right hand	Grip strength (kg) Left hand
11	25	24
12	27	26
13	30	29
14	35	34
15	44	40
16	50	45
17	55	50

Fig 1

Materials

- dynamometer

Method

Use the dynamometer to measure the grip strength (kg) of your right and left hand. Keep your arm straight, without bending it at the elbow, when taking a reading. Record your results.

QUESTIONS

1 (a) Draw a table for recording (i) right-hand and (ii) left-hand grip strengths of all the people in your class. You will need to pool results.
(b) Calculate the mean value for (i) right-hand and (ii) left-hand grip strengths. (6)

2 (a) Subdivide the group into either (i) girls and boys, or (ii) right-handed and left-handed people. Calculate mean values for each group. Present your results in tables. (4)
(b) Write a brief account of your results and conclusions. (4)

3 A dynamometer can be used to diagnose diseases of the muscles and/or nervous system. Name two diseases or conditions in which grip strength may be abnormally low. (2)

4 Look at the table on Sheet 91.
(a) By what percentage is the right hand stronger than the left in boys aged (i) 11 and (ii) 17?
(b) What is the percentage increase in strength of (i) the right hand and (ii) the left hand at 17 compared with that at 11? (4)

Taking it further

1 The following equations relate grip strength to body mass in girls and boys:

For girls:

$$\frac{\text{left-hand grip strength (kg)} + \text{right-hand grip strength (kg)}}{\text{body mass (kg)} \times 150} = 1.0$$

For boys:

$$\frac{\text{left-hand grip strength (kg)} + \text{right-hand grip strength (kg)}}{\text{body mass (kg)} \times 170} = 1.0$$

Figures higher than 1.0 indicate greater than average muscle strength. What are (a) the advantages and (b) the limitations of equations like this?

2 Greater strength in one hand is often associated with greater hand size. Use the following method to find out if one of your hands is bigger than the other.

Stick sheets of graph paper to thick pieces of cardboard (e.g. cereal packets). Cut out 100 cm² of mounted graph paper and weigh it. Record the mass (g). Place your right hand, fingers spread apart, on a mounted sheet of graph paper. Trace around the outline as far as your wrist. Cut out the tracing and weigh it. Record the mass. Divide the mass by the mass of 100 cm² of mounted graph paper, and multiply the figure by 100. This will give the surface area of your right hand. Are both hands the same size, or is one bigger than the other? How does this relate to grip strength?

3 A Bullworker, with a built-in Power-meter, can be used to measure the strength of arm muscles. Try to find out how the strength of these muscles increases over the age range 11–17. How does the strength of arm muscles in girls compare with that in boys?

Investigations in Applied Biology and Biotechnology © 1990 Peter Freeland. Published by Hodder & Stoughton

36 Obesity

Investigation: 40–60 mins

Being overweight has many different causes. In some cases inherited factors, which affect most of the members of a family, are responsible. On the other hand, anyone who eats too much can become obese. Apart from lowering self-respect, **obesity** raises blood pressure and increases the chance of having a heart attack or stroke. People who are obese have to pay more for their life insurance. Also, it is more difficult for obese people to do well in sports and games. They warm up quickly and soon overheat, become short of breath, and tire faster than other people.

One way of finding out if you are overweight is to compare your weight with that of other people of the same height, sex and age. Another method, and one that can show whether you have a tendency to become obese later in life, is to measure skin fold thickness using a pair of skin-fold callipers (Fig 1). Doctors use these to measure the thickness of the triceps skin-fold, about half-way along the back of the upper forearm, or the scapula fold below the shoulder blade (see Fig 2). This investigation compares the thickness of the triceps skin-fold with body mass.

Investigation

Materials

- 10 subjects
- skin-fold callipers
- vertical scale, graduated in cm, for measuring height
- metric scales, for measuring body mass

Method

1 List the names of the subjects. Ask them to remove their shoes and jackets or coats. Measure and record the height (cm) and body mass (kg) of each subject.

2 Identify the triceps skin-fold (Fig 2). Measure and record the thickness of the triceps skin-fold in each subject.

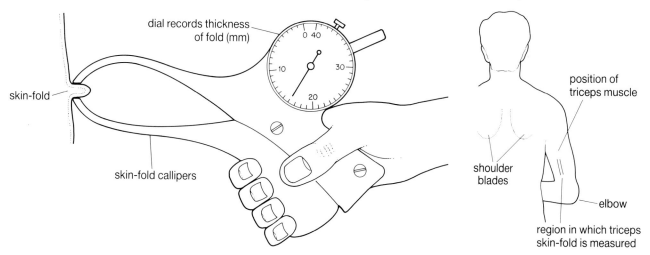

Fig 1   Fig 2

QUESTIONS

1 (a) Copy and complete columns 1 to 4 of the table. (5)

Subject no. and name	Thickness of triceps skin-fold (mm)	Height (cm)	Body mass (kg)	Height (cm)/ body mass (kg)
1				
2				
3				
4				
5				
6				
7				
8				
9				
10				

(b) In general, tall people weigh more than short ones. From the data in columns 3 and 4 devise a simple formula for relating height to body mass. Use your formula to obtain data for the fifth column of the table. (5)
(c) How could the formula be improved? (2)

2 (a) Plot a graph of body mass against triceps skin-fold thickness. (5)
(b) Comment on the relationship. (4)

3 The triceps skin-fold tends to be slightly thicker in girls than in boys, but it shows similar variations in thickness between the ages of 9 and 17 years (see Fig 3). How would you account for
  (a) the changes in the thickness of the skin-fold with age
  (b) the differences between boys and girls? (5)

4 What advice would you give to someone who was obese and who wanted to lose weight? (4)

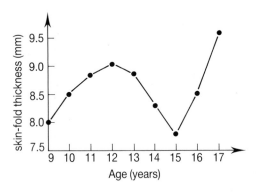

Fig 3 *thickness of the triceps skin-fold: average values for 9–17 year olds*

Taking it further

1 Many young people, especially girls, who believe they are overweight go on a diet. What are the problems associated with dieting? How may they be (a) prevented and (b) overcome?

2 Try to find out whether (a) dieting or (b) exercise has most effect on losing body mass. Design an investigation, lasting for a period of 24 hours, to explore this. Give a full account of the experiment.

37 Successful seed germination

Preparation: 30–40 mins
Investigation: 30–40 mins

Seeds are the simplest, and often the most economical, method of **plant propagation**. Modern methods of producing and packaging seed ensure that a high percentage of the seeds in any packet are **viable**, which means that they can germinate into seedlings. Getting seedlings from seeds, however, depends on the skill and knowledge of the grower. All seeds need moisture, oxygen and a suitable temperature. Others are more demanding, needing particular combinations of light and darkness, softening of the seed coat (**scarification**), or a period of cold treatment (**vernalisation**) before they will germinate. Even when these conditions are right, some seeds may not germinate because they are attacked by fungi or eaten by animals.

In this investigation you will find out how the use of a sterile seed compost and a fungicide can affect the **percentage germination** of the seeds from five different flowering plants.

Preparation

A Sowing in garden soil

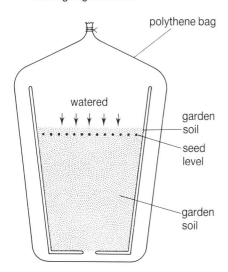

B Sowing in seed compost

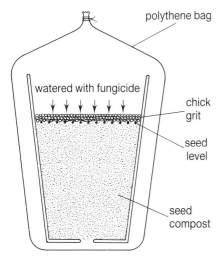

Fig 1

Materials

- seeds of 5 flowering plants (e.g. stock, pansy, polyanthus, arctic poppy, French marigold)
- 200 cm^3 1%$^{w/v}$ Benlate solution
- garden soil
- seed compost
- chick grit
- 10 15–25 cm flower pots
- 10 large plastic bags, e.g. 30 × 40 cm
- spoon
- string
- glass-marking pen
- garden trowel

Method

1 Fill 5 flower pots to within 2.5 cm of the brim with garden soil, and 5 with seed compost. Label the pots.

2 Write the name of one type of seed on each of the soil-filled and compost-filled pots.

3 Open a seed packet. Count out 50 seeds. Sow these, as evenly as possible, over the surface of the soil-filled pot labelled with the same type of seed. Sow 50 seeds of the same type in the compost-filled pot with that label.

4 In the same way, sow each remaining type of seed in turn.

5 Sprinkle a thin layer (3–6 mm) of garden soil over the seeds sown in the soil-filled pots (Fig 1A). Cover the seeds sown on compost with a thin layer of chick grit (Fig 1B).

6 Use the spoon to pour water over the soil-sown seeds until the surface is moist. Water the surface of the compost-sown seeds with Benlate solution (a fungicide).

7 Put each flower pot into a plastic bag, blow up the bag, and tie it with string (Fig 1, Sheet 95). Stand the pots either on shelves in a greenhouse or in artificial light. Leave the pots for 1–2 weeks, or until seedlings are visible.

Investigation

Materials

- 5 flower pots containing soil-grown seedlings
- 5 flower pots containing compost-grown seedlings

Method

1 Count and record the number of seedlings in each pot.

QUESTIONS

1 Copy and complete the table, adding the names of the seeds you have used. (5)

Name of seed	No. of seedlings/pot	
	Garden soil	Seed compost
A		
B		
C		
D		
E		

2 What percentage of each seed type germinated
   (a) in garden soil
   (b) in seed compost? (5)

3 Comment on the general effect of sowing seeds in compost and of watering them with a fungicide. (3)

4 Name:
   (a) three advantages
   (b) one possible disadvantage
   of germinating seeds inside a plastic bag.
   (c) How could the disadvantage be overcome? (5)

5 Suggest two ways in which the design of the investigation could be improved. (2)

Taking it further

1 Devise and carry out an investigation to find out if seeds germinate best in (a) an uncovered plastic pot, (b) an uncovered clay pot or (c) pots of both types, covered by glass or plastic.

2 Some seeds, such as *Lewisia*, *Ginkgo* and oak, need a period of cold treatment before they will germinate. Each month, over a period of a year, sow 20 seeds of one of these species in covered pots, placed outside. Record the percentage germination of seeds sown in each month. What is the best time of year for sowing the seeds?

38 New bulbs from old

Investigation: Part 1 – 30–40 mins
Part 2 – 10–15 mins

Daffodils, hyacinths and tulips are flowering plants grown from bulbs. Plants with bulbs can usually reproduce both asexually and sexually. Asexual reproduction occurs when a mature bulb divides longitudinally to form 1–3 'daughter' bulbs. Sexual reproduction involves pollination and fertilisation, which result in seed production. New plants produced from bulbs are identical to the parent plant. Those which come from seeds usually differ from their parents. So bulb growers produce new varieties from seeds, then increase their stock of these varieties through bulb production. It can, however, take 10–15 years before a grower has enough bulbs of a new variety to sell to the public. Fortunately, biologists have discovered several ways of speeding up bulb production. One of these, described below using daffodil bulbs, can give 4–32 new bulbs from each parent.

Investigation

PART 1

Materials

- daffodil bulb
- 200 cm^3 50%$^{w/v}$ ethanol solution in a 250 cm^3 beaker
- Benlate (a fungicide)
- plastic bag, half-filled with moist vermiculite
- white tile
- forceps
- scissors
- string
- scalpel
- glass rod
- Bunsen burner
- plastic gloves
- eye protection
- access to fume cupboard and greenhouse

Method

1 Put on your eye protection and plastic gloves.

2 Use the scalpel to gently scrape away all the brown-black scaly covering from around the bulb. Do not cut into the soft, white part (Fig 1).

3 Light the Bunsen burner. Flame the scalpel. Use the flamed scalpel to cut the tip and base from the bulb (Fig 2, Sheet 98).

4 Put the cut bulb into ethanol solution. Leave the bulb in ethanol solution for 5 minutes.

5 Use forceps to remove the bulb, allowing as much ethanol solution as possible to drain off. Rest the bulb on the white tile. Flame the scalpel. Cut the bulb lengthways into 4, 8 or 16 segments.

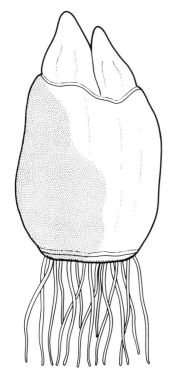

Fig 1

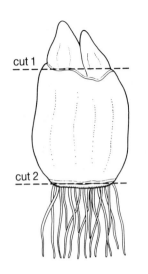

Fig 2

6 Tip one sachet of Benlate (2.25 g) into the 250 cm³ beaker. Half fill the beaker with water. Stir the mixture with a glass rod. Use forceps to put the bulb segments into the Benlate solution. Leave the segments in the fungicide for about 30 minutes.

7 Use forceps to remove the bulb segments from the beaker. Put them into the plastic bag containing moist vermiculite, a moisture-retaining compound. Gently shake the bag until the segments are buried. Tie the bag with string.

8 Store the sealed bag for 10–12 weeks below the shelves of a greenhouse (June–September), or in a warm, dark place.

PART 2

Materials

- plastic bag containing young bulbs
- 2 flower pots
- sterile potting compost, e.g. Levington compost

Method

Pot up the young bulbs in the sterile potting compost, not more than 4–5 per pot (Fig 3).

Fig 3

QUESTIONS

1 Name:
   (a) two flowering plants (1)
   (b) a vegetable (1)
that are propagated from bulbs.

2 In the investigation, why was the cut parent bulb immersed in an ethanol (alcohol) solution? (1)

3 Suppose you bought a bulb of a new daffodil variety for £50 and managed to produce 10 bulbs from it. What would you have to charge for each of these to make a profit of 50%? Show how you arrived at your answer. (3)

4 A commercial grower bought 100 bulbs at £5 each. In the first year 7 new bulbs were produced from each of these. During the second year 5 bulbs were produced from each of the new ones. If all of the bulbs were sold at the end of the second year at £5 each, what was the grower's profit? Show how you arrived at your answer. (4)

Taking it further

It isn't only daffodil bulbs that can be propagated by this technique. You could use the same method to increase the numbers of crocus corms, iris rhizomes, potato tubers and dahlia roots. Design and carry out one or more of the following investigations:

(a) How many new bulbs can you produce from a single large daffodil bulb?

(b) What is the best time of year for propagating daffodil bulbs?

(c) How do different temperatures affect the development of young daffodil bulbs?

(d) How many new potato plants can you produce from a single large potato tuber?

39 New plants from cuttings

Investigation: *Part 1 – 20–30 mins*
(Nov–Dec)
Part 2 – 20–30 mins
(Jan–Feb)
Part 3 – 40–50 mins
(March–April)

Brightly coloured flowering plants, such as geraniums and dahlias, are often expensive. For example, if you went to a garden centre and bought plants that were ready for flowering, each one could cost £1.00 or more. By growing your own plants you can save quite a lot of money, especially if you plan ahead, buy seeds early, and know how to produce more plants from those that you have already. The technique of producing new plants from the leaves, stems or roots of plants is called **vegetative propagation**. Each 'daughter' plant is an exact genetic copy of its parent, identical in shape, size and colour. Vegetative propagation is also a quick way of getting new plants, because you don't need to wait for the seeds to germinate or for young seedling plants to grow from them.

Starting with a single packet of geranium seeds, bought in November, your aim is to produce a bed of 50 or more plants by the following summer. Propagation of the geranium is by stem cuttings, taken in March–April. Dahlias, on the other hand, are propagated by root cuttings, taken at the same time of year. Starting with a single dahlia tuber, your next aim is to subdivide it into 3–6 tubers which will flower in the following summer.

Investigation

PART 1

Materials

- packet of geranium (pelargonium) seeds
- root tuber of dahlia
- Benlate solution (0.1%$^{w/v}$)
- potting/seed compost
- 2 15–25 cm flower pots
- 2 large plastic bags
- string
- spoon
- garden trowel
- natural light, e.g. window sill, or artificial light, e.g. fluorescent lamp

Method

1 Fill both flower pots with compost, to within 5 cm of the rim. Water the compost if it is dry.

2 Use the spoon to flatten the compost in one pot. Sow the geranium seeds evenly over the surface of the compost; press the seeds gently into the compost, but don't cover them.

3 Take 2–3 spoonfuls of Benlate solution and pour it over the seeds to prevent fungi from growing. Put the pot inside a plastic bag. Blow up the bag, then tie it at the neck with string.

4 Plant the dahlia tuber in the other pot, leaving only the top of the stem exposed. Pour 2–3 spoonfuls of Benlate solution over the surface of the compost. Again, put the pot into a plastic bag, blow it up, and tie it at the neck.

Fig 1   geranium seedlings

5 Stand the pots on a windowsill or under a continuously illuminated fluorescent lamp. (A fluorescent lamp can be mounted on a wooden plank, and supported at its ends on bricks.) Leave the pots for 4–8 weeks, until lateral shoots have developed on the geraniums (Fig 1) and buds have sprouted from the dahlia (Fig 2).

PART 2

Materials

- pot containing geranium (pelargonium) seedlings
- pot containing dahlia tuber with buds/shoots
- Benlate solution (0.1%$^{w/v}$)
- potting compost
- 20 15–25 cm flower pots
- 20 plastic bags
- string
- spoon
- garden trowel
- scalpel
- natural light, e.g. windowsill, or artificial light, e.g. fluorescent lamp

Method

1 Pot up each geranium seedling, one per pot. Water each seedling with Benlate solution. Put each pot into a plastic bag. Blow up the bag, tie it, and put it on a windowsill or under a lamp, as before. Leave the pots for 4–8 weeks.

2 Remove the compost from around the dahlia tuber, and use the scalpel to cut the tuber into smaller pieces, each with at least one bud (Fig 2). Separate out the pieces and re-pot them, one per pot. As before, water each root cutting with Benlate solution, and put the pots into plastic bags. Seal the bags and stand the pots in daylight or artificial light. Leave for 4–6 weeks.

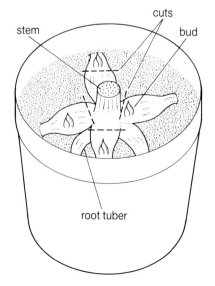

Fig 2   dahlia buds

PART 3

Materials

- potted geranium (pelargonium) plants
- potted dahlia plants
- potting compost
- rooting powder
- 50 15–25 cm flower pots
- forceps
- scalpel
- garden trowel

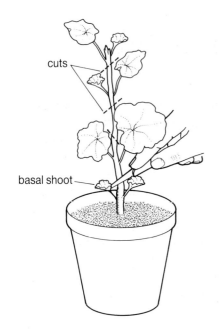

Method

1 Look at Fig 3 which shows how each geranium plant should be cut into segments. Cut off the stem of each geranium above the basal side shoot. (The basal shoot will grow into a new plant.) Lay the cut stem on the bench, and cut carefully between each successive side shoot. Use the forceps to dip the cut stem of each side shoot into rooting powder. Pot up the segments in compost, one per pot. Stand the pots on shelving in a greenhouse. Use the plants for bedding from mid-May.

2 Remove the plastic bags from around the dahlias, and stand the pots on shelving in a greenhouse. Use the plants for bedding from mid-May.

Fig 3

QUESTIONS

1 Name:
 (a) two herbaceous plants (1)
 (b) two shrubs or trees (1)
that are commonly propagated from stem cuttings.

2 At what time of year are most cuttings taken? Suggest a reason for this. (2)

3 What plant hormone is used to encourage root development in cuttings? To which part of the cutting is this hormone applied? (2)

4 Why are some of the leaves removed from a stem cutting before or immediately after it has been planted? (2)

5 What other treatments can be given to a cutting to increase its chance of survival? (2)

Taking it further

1 Plants of *Camellia japonica* can be propagated from their leaves. Cut off a leaf, together with an oval-shaped piece of stem at the leaf-base (Fig 4). Stand the base of the leaf in a $1\%^{w/v}$ Benlate solution for 30 minutes. Pot up several leaf cuttings in a flower pot, using a mixture of 50 parts sharp sand : 50 parts potting compost. Put the pot into a plastic bag, seal the bag, and let it stand in natural or artificial light. Remove the cuttings when they have rooted, and the buds have grown into 5–10 cm shoots.

2 What mass of new potatoes can you grow from a single large seed potato? Plant the original potato in a pot. As soon as buds appear, cut it up into segments. Soak the segments in a $1\%^{w/v}$ Benlate solution for 30 minutes before planting them outside in a row.

Fig 4

40 Rooting cuttings in a gel

Preparation: *20–30 mins*
Investigation: *5–10 mins, over 30 days*
April–September

Softwood (green stem) cuttings can be rooted in soil or almost any medium that retains moisture. One popular artificial medium is a clear **agar-based gel**, which is sold in translucent tubs with a metal foil cover. When you place stem cuttings in this medium, you can observe the various stages of root development and growth. The tubs are intended for kitchen or living-room windowsills. They do not need watering, and their contents will not stain paintwork or carpets if they are accidentally knocked over. As soon as the cuttings have rooted, they can be removed from the gel and potted up in compost.

In this investigation you will prepare your own rooting gels, and find out how adding mineral salts affects the growth of cuttings from two different plants.

Precautions

Stir the molten agar continously while it is being heated, or the beaker may crack. Wear eye protection.

Preparation

Materials

- 0.01 g indole acetic acid (IAA)
- 10 g bacteriological agar powder
- 1 cm³ bleach solution
- ethanol
- 1 g Benlate
- 5 Sach's water culture tablets
- 1 dm³ pyrex beaker
- 500 cm³ pyrex beaker
- 100 cm³ beaker
- test tube fitted with a rubber bung
- 4 translucent plastic cups
- aluminium foil
- scissors
- Bunsen burner, tripod and gauze
- pestle and mortar
- glass rod
- glass-marking pen
- eye protection

Method

1 Put the IAA into the test tube, add a few drops of ethanol, and shake the mixture until a solution has formed.

2 Pour 900 cm³ of water into the 1 dm³ beaker. Add the dissolved IAA, Benlate and bleach solution. Use the glass rod to stir the mixture.

3 Tip the bacteriological agar powder into the 100 cm³ beaker. Add about 70 cm³ of water. Mix to form a slurry.

Investigations in Applied Biology and Biotechnology © 1990 Peter Freeland. Published by Hodder & Stoughton 103

4 Put on your eye protection. Set up the Bunsen burner, tripod and gauze. Light the Bunsen burner. Heat the mixture in the 1 dm³ beaker until it boils. Carefully pour off half the mixture into the 500 cm³ beaker. Re-heat the mixture in the 1 dm³ beaker to boiling point. Slowly add one half of the agar slurry, stirring it with the glass rod. Continue to stir the mixture until the boiling liquid has cleared.

5 Carefully pour the molten agar into two of the cups, to fill them to within 2–3 cm of the rim. Cover the cups with aluminium foil. Allow the agar to cool and harden. Label the cups '– mineral salts'.

6 Crush the Sach's water culture tablets in the mortar. Add the crushed tablets to the mixture in the 500 cm³ beaker. Add the remaining agar slurry. Stir the mixture. Heat the mixture until it boils, stirring continuously. After the mixture has cleared, pour the molten agar into the two remaining cups, to within 2–3 cm of their rims. Cap the cups with aluminium foil. Allow the agar to cool and harden. Label the cups '+ mineral salts'.

Investigation

Materials

- 4 cuttings of geranium (pelargonium)
- 4 cuttings of mint
- 2 agar cups (– mineral salts)
- 2 agar cups (+ mineral salts)
- pencil
- scalpel

Method

1 Trim the geranium cuttings until they consist of two leaves, a terminal bud and approximately 5–10 cm of bare stem (as in Fig 1).

2 Trim the mint cuttings to three pairs of leaves, a terminal bud, and 5–10 cm of bare stem.

3 Use the pencil to make two holes in each aluminium foil cover, spacing the holes about 3 cm apart. Put two cuttings of geranium into one of the agar cups without mineral salts, and two into one of the agar cups with mineral salts (Fig 1). In the same way, set up four cuttings of mint, two in agar with mineral salts, two without.

4 Stand the cuttings in a brightly lit, well ventilated, warm place. At intervals of 2 days over a period of 30 days, count and record the number of (a) roots and (b) new leaves that develop on the cuttings.

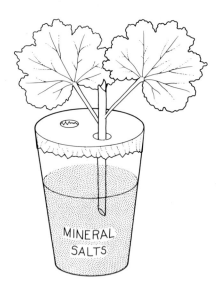

Fig 1

QUESTIONS

1 Copy and complete the table. (Totals are for two plants.) (7)

Day	Geranium		Mint	
	No. roots	**No. new leaves**	**No. roots**	**No. new leaves**
0				
2				
⋮				
30				

2 What were the effects of adding mineral salts to the agar? (4)

3 Name any two problems with the laboratory-prepared gels that you think could be solved. (2)

4 No control was used in this investigation. Suggest a suitable control. (2)

Taking it further

1 Explain the role of each of the following in the gel cups you have prepared: agar; IAA; bleach; Benlate; mineral salts; aluminium foil.

2 A clear gel for rooting cuttings, marketed under the name of Clearcut, is available from garden centres and supermarkets. How does the gel you have prepared in the laboratory compare with this product in terms of (a) cost, (b) appearance and (c) performance? Explain how you assessed each of these points. Suggest ways in which the laboratory-prepared gels might be improved.

Investigations in Applied Biology and Biotechnology © 1990 Peter Freeland. Published by Hodder & Stoughton

41 Micropropagation

Investigation: Part 1 – 50–60 mins
Part 2 – 50–60 mins

Biologists and growers sometimes need to produce large numbers of plants from a single parent. One method they can use is called **micropropagation**. It involves growing plantlets in test tubes, conical flasks or other laboratory glassware from small, sterile pieces of stems, leaves or roots. This technique has great potential for use in the mass production of house plants, plant breeding, and in the transport of plant material over long distances. Furthermore, plants micropropagated from the apical meristems of shoots are usually free from viruses and other pathogens.

In this investigation you will learn the basic principles of micropropagation using root segments of dandelion. When placed in light and kept moist, these segments will develop new shoots and roots without any chemical treatment. After the new shoots and roots have formed the segments can be cut up into smaller pieces. By sub-culturing each of these smaller pieces you can get anything from 500 to 1000 plantlets in less than 8 weeks.

Investigation

PART 1

Materials

- dandelion plant with tap root
- 20 cm^3 1%$^{w/v}$ Benlate solution in a petri dish
- 100 cm^3 beaker with a 2 cm depth of agar gel
- scalpel
- forceps
- Clingfilm
- light source

Method

1 Dig up a dandelion plant. Wash the root under a running tap.

2 Make a cut across the top of the root to remove any leaves and flowers. Use the scalpel to cut the roots into segments, each 0.3–5 cm thick.

3 Keeping the segments top side uppermost, rest them on the bench.

4 Use forceps to put the segments, with the top side uppermost, into the Benlate solution. Let them stand in the Benlate solution for 20–30 minutes.

5 Use forceps to transfer the segments (still with the top side uppermost) onto the surface of the agar in the beaker (Fig 1).

6 Cover the beaker with Clingfilm, and stand it either under a lamp, or on shelves in a greenhouse. Leave the beaker until the segments have developed small green leaves about 0.25–0.5 cm in length (this may take 3–4 weeks).

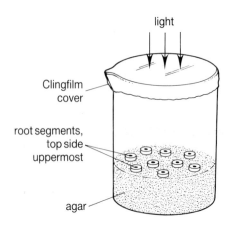

Fig 1

PART 2

Materials

- dandelion segments with leaves
- 20 cm³ 1%^(w/v) Benlate solution in a petri dish
- 4 100 cm³ beakers, each with a 2 cm depth of agar gel
- scalpel
- forceps
- Clingfilm
- light source

Method

1 Use forceps to remove the segments from the beaker. Rest each segment on the bench surface and cut it lengthways into 8 pieces (Fig 2).

2 Use forceps to transfer the cut pieces into Benlate solution. Leave them in the Benlate solution for 20–30 minutes.

3 Use forceps to place the smaller pieces on the surface of the agar in the beakers (Fig 3). Cover each beaker with Clingfilm. Leave the beakers under a lamp or on shelves in a greenhouse until they have developed shoots and roots (this could take 3–4 weeks).

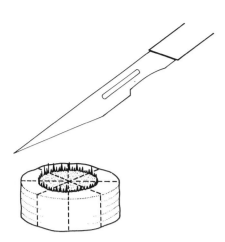

Fig 2

Fig 3

QUESTIONS

1 What are the commercial advantages of micropropagation? Does micropropagation have any disadvantages? (3)

2 What is the function of each of the following chemical compounds, used in micropropagation:
  (a) auxin
  (b) gibberellin
  (c) cytokinin
  (d) agar
  (e) sodium hypochlorite? (5)

3 What do you understand by the term 'clone'? (2)

Taking it further

1 Devise and carry out investigations to find answers to one or more of the following questions:

(a) Do thick dandelion root segments grow shoots sooner than thin segments?

(b) Do segments cut from the top of a root grow shoots sooner than those cut from lower down?

(c) What is the minimum width of a segment that will grow shoots?

(d) Do root segments of (i) dock and (ii) bindweed develop shoots as readily as those of dandelion? If so, what is the significance of this for farmers and gardeners?

2 Try to produce carnation plants from apical buds, cut from a mature plant in May–August. Set up sterilised buds in agar gels containing 2.0, 1.0, 0.5 and 0.1 mg cytokinin (BAP)/dm³ solution. After continuous illumination for 2–3 weeks, or as soon as the buds have produced shoots of 1–2 cm, transfer them to agar gels containing 1.0, 0.5 and 0.1 mg auxin (IBA)/dm³ solution. Further continuous illumination should induce roots to form.

42 Pollination

Investigation: Part 1 – 10–15 mins
(June–July)
Part 2 – 10–15 mins
(Dec–Jan)
Part 3 – 20–30 mins
(March–April)
Part 4 – 10–15 mins
(June–July)

New varieties of plants are produced from seeds. Plant breeders select parent plants with different features which they want to combine, such as red flowers and long stems, and cross-pollinate them in the hope that some of the plants grown from the resulting seeds will have the combined features. As a result of carefully controlled programmes of **cross-pollination**, many different varieties of each species are now available.

In this investigation you will learn the basic principles of controlled cross-pollination using polyanthus or primrose plants grown in pots. The individual plants are of two types: having flowers that are either **pin-eyed** or **thrum-eyed** (Fig 1). In pin-eyed flowers, the end of a long **style** capped by a rounded **stigma** can be seen at the centre. The **stamens** of these flowers are half-way down the pollen tube, hidden from view. Thrum-eyed flowers have a ring of stamens at their centre. The stigmas of these flowers can't be seen from the outside, because the styles are short. Crosses can be made only between pin-eyed and thrum-eyed plants, as the pollen will not grow on the stigma of the flower in which it is produced.

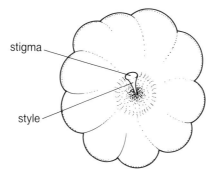 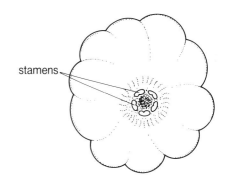

Fig 1

Investigation

PART 1

Materials

- polyanthus or primrose seeds in 2 different varieties/strains, A and B
- 2 seed boxes
- seed compost
- garden trowel

Method

Sow the seeds in boxes, keeping the compost moist and shaded.

PART 2

Materials

- polyanthus or primrose plants in 2 varieties/strains, A and B
- Benlate solution (1%$^{w/v}$)
- potting compost
- 10 15–25 cm flower pots
- 10 plastic bags
- string
- spoon
- garden trowel

Method

Pot up 5 plants of each variety/strain using potting compost. Use the spoon to water the plants with the Benlate solution before sealing each plant in a blown-up plastic bag.

PART 3

Materials

- pin-eyed flowering plant of variety/strain A
- thrum-eyed flowering plant of variety/strain B
- scissors
- plastic bag
- string

Method

Wait until the plants have 6 or more flowers.

1 Select a pin-eyed plant of one variety/strain and a thrum-eyed plant of the other. Remove the plastic bag. Cut off the thrum-eyed flowers. Using scissors, trim back the petals of these flowers to expose the stamens (see Fig 2).

2 Take a trimmed thrum-eyed flower between your thumb and index finger. Rub pollen from its stamens over the stigma of a pin-eyed flower (Fig 2).

3 Put the pin-eyed plant into a plastic bag, blow up the bag and seal it with string. Stand the pot in a greenhouse, preferably under shelving.

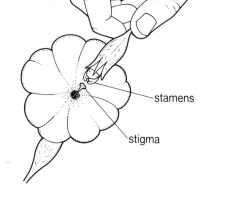

Fig 2

PART 4

Materials

- pollinated plant bearing seed pods

Method

1 Harvest the seeds.

2 Sow the seeds immediately after harvest.

3 Pot up seedlings in November–January.

4 Select the most suitable plants when flowers appear (March–April).

QUESTIONS

1 Name two different animal agents that could cross-pollinate polyanthus or primrose plants. (2)

2 In the investigation, what were the reasons for doing each of the following?

(a) Keeping the plants in plastic bags before and after pollination.

(b) Removing petals from the thrum-eyed flower used as a source of pollen. (2)

3 What would you conclude if a cross between a red-flowered and white-flowered plant produced

(a) all white-flowered plants

(b) all red-flowered plants

(c) all pink-flowered plants

(d) red-flowered and white-flowered plants in roughly equal numbers? (4)

4 Plants with pin-eyed flowers have the genotype ss, while those with thrum-eyed flowers are heterozygotes, Ss. What will be the proportions of pin-eyed and thrum-eyed plants in the next generation? Show how you arrived at your answer. (2)

Taking it further

Try to cross-pollinate varieties/strains of any of the following potted plants:

Amaryllis; daffodil; heather; hyacinth; iris; *Lewisia*; *Lychnis*; snowdrop; tulip.

As self- or insect-pollination can occur in most of these plants, you will need to take the following precautions:

(a) Cut out the stamens from young flowers of one variety/strain before self-pollination can take place.

(b) Use mature flowers of the other variety/strain as a source of pollen.

(c) Seal pollinated flowers in polythene bags to prevent other pollen grains from reaching them.

Investigations in Applied Biology and Biotechnology © 1990 Peter Freeland. Published by Hodder & Stoughton

43 Longer life for cut flowers

Preparation: 15–20 mins
Investigation: 30–40 mins, over 5–9 days

Cut flowers are very attractive. However they are usually quite expensive, and may wither and die only a few days after being put into a vase of water. You may wonder why a cut flower never lasts as long as one growing outside that is still attached to the plant. No one really knows the answer to that question, but there are several theories which you can test by a simple investigation. Biologists believe that the life of a cut flower may be shortened by one or more of the following factors:

(a) the stem is unable to absorb water
(b) the flower runs short of nutrients, such as sucrose
(c) bacteria, protozoa and algae grow in the water, blocking xylem vessels
(d) the flower becomes stressed because of a shortage of plant growth substances, including asprin
(e) room temperatures are too high.

You are asked to investigate how some of these factors affect the **vase-life** of daffodils (or some other flower) picked in the bud stage (Fig 1).

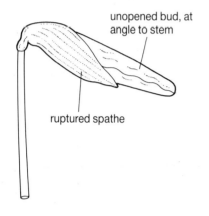

Fig 1   daffodil flower: bud stage

Preparation

Materials

- 50 daffodils or other flowers (freshly picked)
- 5 500 cm³ conical flasks
- 50 g sucrose
- 0.5 g sodium salicylate
- packet of commercial flower preservative, e.g. Chrysal
- hammer
- glass rod
- glass-marking pen

Method

1 Number the conical flasks 1–5. Pour 400 cm³ of tap water into each flask. Put 10 flowers into flask 1. Label the flask 'intact stems'.

2 Take another 10 flowers. Use the hammer to crush the ends of the flower stalks. Put them into flask 2, and label the flask 'crushed stems'.

3 Add the sucrose to flask 3. Stir to dissolve. Put 10 flowers into the flask. Label the flask 'sucrose'.

4 Add the sodium salicylate to flask 4. Stir to dissolve. Put 10 flowers into the flask. Label the flask 'asprin'.

5 Prepare the commercial flower preservative according to the manufacturer's instructions. Pour the preparation into flask 5. Put 10 flowers into the flask. Label the flask 'flower preservative'.

6 Stand all the flasks together on the bench, preferably at a temperature of 12–18°C.

Investigation

Keep a daily record of the number of dead flowers (Fig 2) per flask, until all the flowers have died.

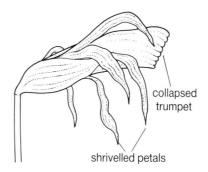

Fig 2 dead daffodil flower

QUESTIONS

1 Copy and complete the table. (8)

Flask no.	No. living flowers/flask										
	Day 0	1	2	3	4	5	6	7	8	9
1										
2										
3										
4										
5										

2 Plot a graph to show how the most effective treatment prolonged the vase-life of the cut flowers. (4)

3 Crushing the end of the stem allowed more water to reach the flowers.
 (a) What was its effect on vase-life? (1)
 (b) How would you find out (i) whether the xylem vessels of flowers in flask 1 were blocked, and (ii) whether the water contained bacteria, algae and protozoa? (2)

4 Comment on the effects of
 (a) sucrose
 (b) sodium salicylate
 (c) the commercial flower preservative. (3)

5 What treatment(s) would you suggest to anyone who wanted their cut flowers to last longer? (2)

Taking it further

1 The following species are widely sold as cut flowers: daffodil, tulip, iris, carnation, sweet pea, chrysanthemum. Design and carry out an investigation to find out which one has the longest vase-life. Which provides the best value for money?

2 Design, write and carry out an investigation to find the effects of temperature on the vase-life of cut daffodils or some other flower.

3 Anti-transpirants are sold to stop the needles falling from Christmas trees. Do anti-transpirants sprayed over the petals of cut flowers extend their vase-life? What about antiperspirants, sold as body sprays? Do they have any effect on the vase-life of cut flowers? If so, when is the best time to spray the flowers?

44 Dispersing weed seeds

Investigation: *40–60 mins*

A **weed** is an unwanted plant that grows on cultivated soil, often between flowers, vegetables or cereals. Weeds don't just spoil the appearance of cultivated plants, they compete with them for light, water and mineral salts. As a result, weed-infested plants are smaller, with fewer leaves, flowers and fruits, than those that are weeded regularly. Furthermore, many weeds carry pests, such as aphids or fungi, that may spread to attack the other plants.

Weeds have many characteristics. For example, they grow rapidly, they have short life cycles and they produce either many daughter plants or large numbers of seeds. These features combined help weeds to spread quickly to areas of freshly dug soil, such as gardens or ploughed fields. As weeds can damage crops, weed control is an important area of applied biology. It is essential to remove weeds before they produce fruits, otherwise their seeds will germinate in the soil to produce many more weeds in the following season. This investigation deals mainly will annual seed production and dispersal in dandelion. Note that a dandelion fruit contains a single seed (see Fig 1).

Fig 1 *fruiting head of a dandelion - each fruit contains a single seed*

Investigation

Materials

- ripe fruiting head of dandelion with 10 or more fruits
- metre rule
- scalpel
- magnifying glass
- retort stand, boss and clamp
- stop-clock, or watch with a second hand

Method

1 Remove all the fruits from the seed head and put them into the petri dish.

2 Look at the top of the receptacle which the fruits have been removed from. You will notice that the surface is pitted. Each pit marks the position of a single fruit.

QUESTIONS

1 Examine the receptacle which the fruits have been removed from. Devise and use a quick method for finding the approximate number of fruits in the complete head. Explain how you arrived at your answer. (4)

2 A large dandelion plant produced 8 fruiting heads in May and a further 4 in September. What was the approximate number of fruits (seeds) produced in a year? Show how you arrived at your answer. (2)

3 Make a large drawing ($\times 5-7$) of a single dandelion fruit. (3)

4 Above the fruit is a structure like a parachute, formed from a ring of hairs.
 (a) Devise and carry out an experiment to find the effect of the parachute on the rate at which the fruit falls. Fruits could, for example, after suitable treatments, be dropped from a fixed height above the bench. Describe your method. (4)
 (b) Present your results in the form of a table. (4)

5 (a) Name one environmental factor that could affect your results. (1)
 (b) What are the most likely functions of (i) the parachute, and (ii) minute hooks surrounding the fruit? (2)

Taking it further

1 Ask your teacher for help with the identification of garden weeds. Write down the methods by which each weed species spreads, and any other features that make it successful in competing with cultivated plants.

2 Many plants have wind-dispersed seeds. Investigate the rate at which other parachute types (poplars, willow herbs, goat's beard, old man's beard, for example) travel in still air.

3 Single fruits of sycamore spin like the blades of a helicopter as they fall. How does the length of sycamore fruits affect the rate at which they fall? At what speed do the fruits rotate?

Investigations in Applied Biology and Biotechnology © 1990 Peter Freeland. Published by Hodder & Stoughton

45 Trapping and killing slugs

Preparation: Part 1 – 10–15 mins
Part 2 – 20–30 mins
Investigation: 40–50 mins, over 3–4 days
March–July

Precautions

Slug pellets are poisonous.

There are more than 20 different kinds of slugs in the UK. Most slugs feed on young plants during the night. Because of this, and because there are often so many of them, slugs can cause severe damage to garden plants, mainly vegetables and flowers. This can be a serious problem for commercial growers.

One method of trapping slugs is to cover the soil with squares of carpet or newspaper. The slugs cling to the moist underside of the material, making them easy to collect and destroy. Alternatively, the surface of the soil can be sprinkled with **slug pellets**, usually made from a mixture of bran and **metaldehyde** (Fig 1). The bran attracts the slugs, which feed on the mixture. The metaldehyde then poisons them and causes increased slime production. The slugs remain on the soil surface surrounded by thick slime, and usually die from drying out (desiccation), especially during dry weather.

In this investigation you will compare the effectiveness of the two trapping methods, and find out how thickly you need to apply slug pellets for them to act as an effective bait.

Preparation

PART 1

Materials

- slug pellets
- 5 small plastic bags
- plastic gloves
- top-pan balance
- glass-marking pen

Fig 1

116 Investigations in Applied Biology and Biotechnology © 1990 Peter Freeland. Published by Hodder & Stoughton

Method

Put on your plastic gloves. Weigh out the following masses of slug pellets: 1 g, 5 g, 10 g, 15 g and 20 g. Put each batch into a plastic bag. Label the bags.

PART 2

Materials

- slug pellets (in plastic bags)
- 5 1 m² squares of carpet (or 10 0.5 m² squares of newspaper)
- 1 m² quadrat
- 5 bricks or large stones (10 if newspaper is used)
- 20 sticks
- hammer
- plastic gloves
- cultivated soil, e.g. garden soil

Method

1 Select an area of cultivated soil where slugs are present. Lay the squares of carpet or newspaper on the soil. Place a brick or large stone on each square.

2 Put on your plastic gloves. Lay the quadrat on the soil surface. Carefully and evenly sprinkle the slug pellets from the 1 g bag over the area inside the quadrat. Place sticks at the corners of the quadrat to mark its position.

3 In the same way, scatter each of the other bags of slug pellets over a 1 m² area of the soil. Mark the position of each quadrat and draw a sketch map of the area.

4 Each morning over a period of 3 days, count and record the number of slugs (a) trapped beneath the carpet/newspaper and (b) trapped by the slug pellets. Remove and destroy all trapped slugs after counts have been made.

QUESTIONS

1 Copy and complete the tables.
  (a) carpet/newspaper traps   (6)

Day	No. slugs trapped					Total catch
	Trap no. 1	2	3	4	5	
1						
2						
3						
Totals						

Investigations in Applied Biology and Biotechnology © 1990 Peter Freeland. Published by Hodder & Stoughton

(b) slug pellet traps (6)

Day	No. slugs trapped					Total catch
	Bait 1 g	5 g	10 g	15 g	20 g	
1						
2						
3						
Totals						

2 What conclusions can you draw from your results? (3)

3 Criticise the design of the experiment and suggest improvements. (3)

4 Slug pellets are often coloured with a blue dye. Suggest two reasons for this. (2)

Taking it further

1 Find out more about the natural predators of slugs, such as hedgehogs, frogs and large ground beetles. Make a list of all the animals that feed on slugs. What conditions should you create in your garden to (a) discourage slugs and (b) encourage the animals that eat slugs?

2 Slugs eat many different plants, but they prefer some plants to others. Design and carry out an investigation to find out what food slugs like best. Use three or more of the following vegetables: lettuce; cabbage; turnip; swede; parsnip; onion; carrot; beetroot; cucumber; marrow; pea; potato; celery; tomato. Are all the parts of a flowering plant, such as a dahlia, equally likely to be attacked by slugs?

3 In addition to bran, slugs are also attracted to beer. If you fill beakers or similar containers with beer, and then bury them so that the top of the beer is just below surface level, the slugs fall in and drown. Design and carry out an investigation to find out if light ale, brown ale or lager is most effective as a slug bait.

46 Soil conditioning

Investigation: *40–60 mins*

Good garden soil is a complex mixture of minerals, organic matter, air, water and living organisms. If your garden contains only mineral particles, such as sand or clay, it may be difficult to grow good quality flowers and vegetables. Sandy soil becomes powdery and loose in summer, allowing water and nutrients to drain through it quickly. Clay soils aren't much better. They become wet and sticky in winter, making them cold and difficult to dig. Most gardeners would like a soil that is neither of these.

Acta-bacta is a preparation of ground **lignite**, an intermediate compound in coal formation, and is sold as a soil conditioner. The manufacturer claims that it has the following effects:
(a) increases the water-holding capacity of sands and gravels
(b) rapidly improves the structure of heavy clays
(c) improves nutrient economy by preventing **leaching**
(d) increases phosphate availability in the presence of lime.

In this investigation you will find out how Acta-bacta affects the **porosity** (drainage) of sand and clay soils.

Investigation

Materials

- 70 g dry sand
- 70 g dry clay
- 10 g Acta-bacta
- 2 100 cm^3 beakers, graduated in 20 cm^3 (or 25 cm^1) units
- 50 cm^3 (or 100 cm^3) measuring cylinder
- filter funnel
- teaspoon
- 4 filter papers
- stop-clock, or watch with a second hand

Method

1 Set up the filter funnel, as shown in Fig 1, supported by the rim of the measuring cylinder. Fold one of the filter papers and fit it inside the filter funnel.

2 Take one of the 100 cm^3 beakers and fill it to the 10 cm^3 mark (or 25 cm^3 mark) with dry sand. Tip the sand into the filter funnel.

3 Pour 50 cm^3 of water into the clean 100 cm^3 beaker. Tip the water over the sand. Set the stop-clock, or record the time on your watch. Record the volume of water in the measuring cylinder every 30 seconds, until water has stopped dripping from the sand.

4 Measure out the same volume of sand (20 cm^3 or 25 cm^3) in the dry beaker. Add a level teaspoon of Acta-bacta. Mix thoroughly. Repeat instructions 2 and 3, recording the rate at which water collects in the measuring cylinder.

5 Repeat instructions 2, 3 and 4 using clay. Record all your results.

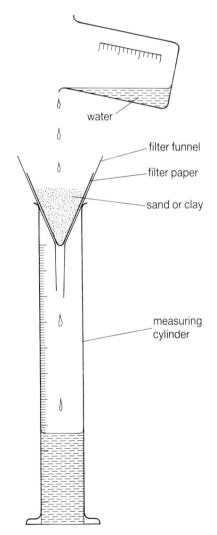

Fig 1

QUESTIONS

1 Copy and complete the table. (6)

Time (s)	Vol. water collected in measuring cylinder (cm³)			
	Sand	**Sand + Acta-bacta**	**Clay**	**Clay + Acta-bacta**
0				
30				
60				
90				
⋮				
570				
600				

2 What was the volume of water retained by
  (a) sand
  (b) sand + Acta-bacta
  (c) clay
  (d) clay + Acta-bacta? (4)

3 Calculate the percentage water retained by
  (a) sand
  (b) sand + Acta-bacta
  (c) clay
  (d) clay + Acta-bacta
Show how you arrived at your answers. (4)

4 Plot graphs to show the effect of Acta-bacta on
  (a) sand
  (b) clay. (8)

5 What do you conclude about the effects of Acta-bacta on
  (a) sand
  (b) clay? (2)

6 List three faults in the design of the investigation. Suggest three improvements. (6)

Taking it further

1 The manufacturer recommends that Acta-bacta should be applied to the surface of soil at the rate of approximately 1 kg/m².

Select two adjacent garden plots. Dig both, then treat the surface of one plot with Acta-bacta. Plant the same number of onions, potatoes, cabbages, peas, beans or other vegetables in each plot. Keep a diary to record the height of plants in each plot. In addition, record any attacks by diseases and insects. At harvest, find the mass of each crop. Tabulate your results.

What were the effects of adding Acta-bacta to the soil? Was the market value of any extra vegetables produced greater than the cost of the treatment?

2 Design and carry out an experiment to find out if other materials, such as peat, compost and lime, are useful as soil conditioners.

47 Composting

Preparation: 20–30 mins
Investigation: 30–40 mins, over 5–10 days
April–September

Organic fertilisers, such as farmyard manure and composted vegetation, are excellent soil conditioners. They help to improve soil texture in a number of ways, mainly by retaining moisture and increasing the amount of air present. Furthermore, they slowly release mineral salts into the soil, which promotes better growth of flowers, fruits and vegetables.

Household waste, including hedge trimmings, lawn mowings and vegetable scraps, often find their way into the dustbin. An alternative method of disposal is **composting**, in which the end product can be used in the garden to improve both the soil and the plants growing in it. All vegetation can be composted, but grass mowings are particularly suitable because they break down quickly when piled into heaps.

When grass mowings are decomposing they produce a lot of heat from bacterial activity. This investigation looks at the temperature changes during composting, with and without the addition of compounds that provide food for bacteria and fungi.

Preparation

Materials

- grass mowings (freshly cut)
- large freezer bag (2–3 kg capacity)
- polythene carrier bag (10–15 kg capacity)
- 2 polythene sacks (50 kg capacity)
- 1 kg sucrose
- commercial compost-maker (e.g. Garotta)
- thermometer
- string
- garden fork
- watering can giving a fine spray

Method

1 Take the freezer bag, polythene carrier bag and one of the sacks. Fill each to the brim with grass mowings. Compress the grass. Tie the neck of each container with string.

2 Select your mowings for the second sack. Use the garden fork to mix these with sucrose, an energy source for micro-organisms, and compost-maker, a source of mineral salts. Add enough water to make the mixture moist. Fill the sack with the mixture, compress it and tie the neck of the sack.

3 Measure and record the air temperature and the temperature at the centre of each container.

Investigation

Materials

- containers filled with grass mowings

Method

1 Each day over a period of 5–10 days, measure and record the air temperature and the temperature at the centre of each container. Record your results.

2 At the end of the investigation inspect the mowings at the centre of each container. Record any differences.

QUESTIONS

1 Draw a table to show all your results. (4)

2 Plot a graph to show variations in
(a) the air temperature
(b) the temperature at the centre of each container. (6)

3 What do you conclude from your results? (3)

4 (a) How do grass mowings in the freezer bag compare with those in the sacks?
(b) Try to explain any differences. (2)

Taking it further

1 Compost can be made by each of the three methods shown in Fig 1. List the advantages and disadvantages of each method. Collect well-rotted material from a compost heap in March–April. Put this material into polythene bags, tie them, and keep the bags sealed until June–July. Open the contents of the bags into a Tullgren funnel. Collect, count and identify any animals that emerge from the compost. How might these animals be used to speed up the process of composting?

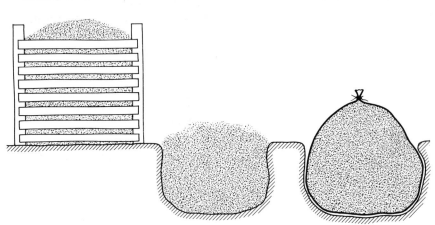

A Wooden frame, above ground B Open soil pit C Sealed black polythene bag

Fig 1

2 Brandling worms, which live in compost and manure heaps or can be bought from angling shops, will convert kitchen scraps into a high quality seed compost. Make your wormery from a large polythene bucket fitted with a lid and a circular sieve (see Fig 2). Remove the bottom of the bucket using sharp scissors. Fit the sieve inside. Fill the bucket with finely chopped kitchen scraps. Add 10–20 worms. Replace the lid and stand the bucket on a brick support. Remove any compost that collects beneath the bucket, and add more kitchen scraps as compost forms.

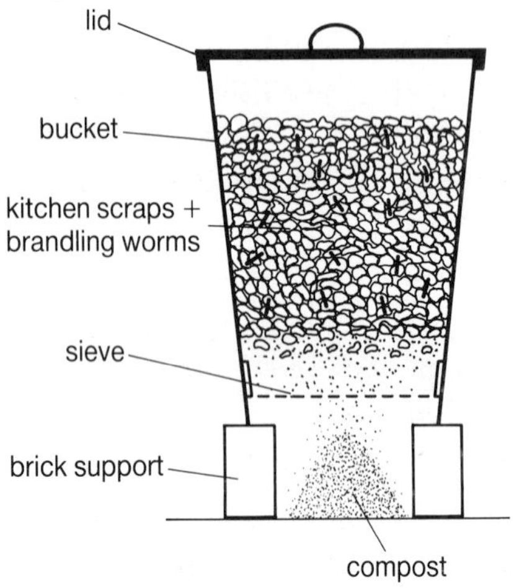

Fig 2

48 Making a biogas generator

Preparation: 60–80 mins

Precautions

Wear eye protection. Heat/burn the plastic container only in a fume cupboard. Avoid contact with Plastic Padding. Do not attempt to ignite biogas.

Preparation

Biogas is a mixture of methane, carbon dioxide and hydrogen sulphide produced by the anaerobic digestion of farm, factory or kitchen wastes. The gas, which is highly inflammable, can be used for lighting and cooking. It can also be used for generating electricity, or it can be compressed into cylinders for use in cars or tractors. Separating biogas from its raw materials is not a problem, because the gas simply bubbles off and can be collected. Afterwards, the raw material, now in the form of a black slurry or sludge, is usually shovelled out and used as an organic fertiliser.

In this investigation you will make a biogas generator from a 5 dm^3 plastic container and a length of rubber tubing. Using kitchen waste as the raw material, biogas samples will be collected over water in a gas jar.

Materials

- 2–3 kg kitchen waste
- 2–3 dm^3 rain water
- 5 dm^3 plastic container
- 5 cm^3 gas jar
- beehive shelf
- rubber tubing, 1.5–2.0 m × 1.5–2.5 cm diameter
- plastic bucket
- kitchen mincer or blender
- bradawl, wooden handled
- paintbrush
- aluminium paint
- Plastic Padding
- Bunsen burner
- scalpel
- eye protection
- access to fume cupboard

Method

1 Put on your eye protection. Set up the Bunsen burner in a fume cupboard. Light the burner and turn on the fume cupboard extractor fan. Heat the metal part of the bradawl until it is red hot. Burn a circular hole, of approximately the same diameter as the rubber tubing, in one side of the plastic container, 2–3 cm from the top.

2 Fit the rubber tubing as shown in Fig 1, then seal the space between the container and tubing with aluminium paint. Allow the paint to dry and harden (30–60 minutes).

3 Mix Plastic Padding according to the manufacturer's instructions. **Care: Avoid skin contact with the hardener.** Use the scalpel to apply Plastic Padding to the surface of the container, sealing around the rubber tubing. Allow 15–20 minutes for the Plastic Padding to dry and harden.

Fig 1

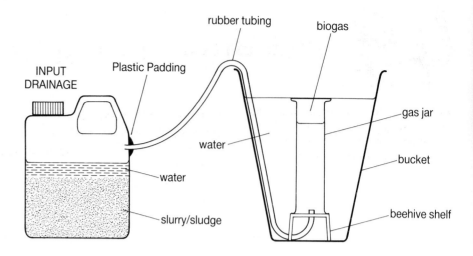

Fig 2

4 Take the kitchen waste and pass it through the kitchen mincer or blender to reduce the size of the particles. Half fill the container with this processed kitchen waste, then cover it with rain water, gently shaking the mixture to produce a slurry/sludge. Replace the screw top.

5 Half fill the bucket with water. Invert the gas jar, filled with water, over the beehive shelf. Insert the end of the rubber tubing through the aperture in the beehive shelf, into the base of the upturned gas jar. Leave the apparatus, shown in Fig 2, until some biogas has collected in the jar. (This could take 1–4 weeks.)

QUESTIONS

1 A farmer bought a biogas generator for his farm. What raw waste materials could he put into it? (2)

2 Biogas generators used on farms are often built below ground level. Suggest two reasons for this. (2)

3 Why should a biogas generator never be built close to a house? (3)

4 Explain the importance of biogas generators in countries such as China and India. (3)

Taking it further

Starting with two plastic dustbins, design and build a biogas generator suitable for processing waste from your kitchen. The generator will need an input point for refuse, a drainage point for removing sludge, a container for gas storage and a tap for regulating gas flow. Plastic pipes or rubber tubing may be used to connect the two dustbins. Plastic Padding, applied over a layer of aluminium paint, will produce water-tight seals.
(Molecules of methane (CH_4) are less dense than molecules of oxygen (O_2) or nitrogen (N_2) and so they will collect at the top of your gas storage tank.)

49 The effects of acid rain

Preparation: *15–20 mins*
Investigation: *30–50 mins, over 4–5 days*

Burning coal and oil releases toxic gases into the air. Winds may carry these gases, mostly nitrogen and sulphur, high into the atmosphere. Here, nitrates and sulphates are formed, with ozone and sunlight probably acting as catalysts. When moisture in clouds precipitates as rain, these compounds dissolve in the water to form nitric and sulphuric acids, the major components of **acid rain**.

Growing evidence suggests that acid rain has widespread damaging effects on plants and animals. It may, for example, be responsible for killing forest trees and aquatic plants and animals. Humans may be most at risk from aluminium compounds. These are released from the soil by acid rain, drain into rivers and then enter the domestic water supply.

Acid rain may also affect the germination of some seeds. This could have important economic implications for farmers and growers. This is an investigation of the effects of different dilutions of nitric (HNO_3) and sulphuric (H_2SO_4) acids on the germination of mustard seeds.

Preparation

Materials

- mustard seeds
- 10 cm^3 0.1 M nitric acid in a suitable container
- 10 cm^3 0.1 M sulphuric acid in a suitable container
- 100 cm^3 distilled water in a beaker
- 9 petri dishes
- 6 flat-bottomed tubes
- 9 filter papers
- 2 10 cm^3 plastic syringes
- scissors
- glass-marking pen

Method

1 Take the petri dishes and cover the base of each dish with filter paper. Trim the papers with scissors until they fit.

2 Use one of the syringes to transfer 2 cm^3 of 0.1 M nitric acid to one of the petri dishes. Label it '0.1 M nitric acid'.

3 Set up three flat-bottomed tubes, labelled 1–3. Put 1 cm^3 of 0.1 M nitric acid into tube 1. Add 9 cm^3 of distilled water to it. Pipette 2 cm^3 of the solution into a petri dish and label it '0.01 M nitric acid'.

4 Transfer 1 cm^3 of the mixture from tube 1 into tube 2. Add 9 cm^3 water. Transfer 2 cm^3 of this solution into a petri dish and label it '0.001 M nitric acid'.

5 Make a further dilution of the nitric acid to give a solution of 0.0001 M. Transfer 2 cm^3 of this solution into a petri dish and label.

Investigations in Applied Biology and Biotechnology © 1990 Peter Freeland. Published by Hodder & Stoughton

125

6 In the same way, set up petri dishes containing 2 cm³ of 0.1, 0.01, 0.001 and 0.0001 M sulphuric acid. (Use your second syringe when working with sulphuric acid.)

7 Transfer 2 cm³ of water to the remaining petri dish. Label it.

8 Sow mustard seeds in each petri dish, 5 or 10 seeds per dish. Replace the lids of the dishes. Allow 3–5 days for germination.

Investigation

Materials

- petri dishes containing germinating mustard seeds

Method

Measure and record the mean (average) length of the radicles in each dish (Fig 1) after those in water have reached a length of 3–5 cm.

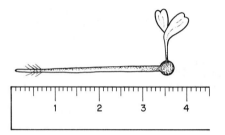

Fig 1

QUESTIONS

1 Present your results in the form of a table. (4)

2 Draw a graph of your results. (6)

3 What do you conclude about
 (a) the general effect of acid (1)
 (b) the specific effects of HNO₃ and H₂SO₄ on the germination of mustard seeds? (2)

4 Try to account for your results. (2)

Taking it further

1 Extend the investigation to include the effects of (a) hydrochloric acid, (b) sodium hydroxide and (c) sodium chloride on the germination of mustard seeds. Is the germination of other seeds influenced to the same extent? What effect, if any, do these compounds have on the growth of the plumule?

2 Collect water samples from local ponds, lakes and rivers. make accurate measurements of the pH using a pH meter. Does the pH change at different seasons of the year? Is there any evidence that the water is becoming more acid?

50 Water purification

Investigation: *40–60 mins*

Clean drinking water is essential for good health. In the UK drinking water comes from lakes, reservoirs and rivers. Much of this water is unfit for human consumption because it is polluted by effluents from farmland. This effluent contains mineral salts, mostly nitrates, that encourage the growth of bacteria, algae, fungi and protozoa. These organisms, especially those which can cause disease, must be removed before the water is fit to drink. Filtration is the first stage in water purification. Water is allowed to filter through beds of sand and gravel. As a result microbes are filtered out, together with fine particles of mud.

The main aim of this investigation is to illustrate the principles of **water filtration**. You will also find out how the arrangement of different sized mineral particles – clay, sand, gravel – affects the rate of filtration and the clarity of the filtered water.

Investigation

Materials

- 10 g clay
- 10 g sand
- 10 g gravel
- 2 20 cm^3 plastic syringes
- 10 cm^3 plastic syringe
- 2 retort stands, bosses and clamps
- 2 10 cm^3 measuring cylinders
- teaspoon
- stop-clock, or watch with a second hand
- glass-marking pen

Method

1 Label the 20 cm^3 syringes A and B. Use the spoon to put clay, sand and gravel into the plastic syringes in the following order, and in these quantities:

	Syringe A	Syringe B
(a)	4 cm^3 gravel	4 cm^3 clay
(b)	4 cm^3 sand	4 cm^3 sand
(c)	4 cm^3 clay	4 cm^3 gravel

2 Tap the base of each syringe to remove any large air spaces. Clamp each syringe with the nozzle 10–15 cm above the bench. Put a measuring cylinder under each syringe (Fig 1).

3 Use the 10 cm^3 syringe to put 3–4 cm^3 of water into each syringe. Allow the water to drain through the layers of mineral particles. After drainage is complete, empty the measuring cylinders, and then return them to their positions.

4 Start the stop-clock, or record the time on your watch. Add 10 cm^3 of water to each 20 cm^3 syringe. At intervals of 5 minutes measure and record the volume of filtered water in each measuring cylinder. Take readings until 9 cm^3 of water has collected in each cylinder.

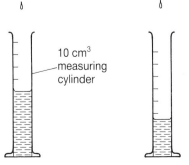

Fig 1

QUESTIONS

1 Present your results in the form of a table. (5)

2 Draw a graph of your results. (5)

3 In which syringe was the water
  (a) cleanest
  (b) filtered most rapidly? (2)

4 (a) Which arrangement of mineral particles would you recommend for use in a filter bed? Give reasons for your answer. (4)

(b) How could you find out if filtration had removed all the bacteria, algae and protozoa from a water sample? (2)

(c) What further treatment(s) would you give to filtered water before it was safe for humans to drink? (2)

Taking it further

1 Design and draw a large-scale plant for water filtration. Your design should include a continuous feed into the plant, together with storage space for the water that has been filtered.

2 Filter beds, consisting mainly of sand, with layers of immobilised enzymes bound to the surface of a gel, can be used to clarify beer, wines and spirits.

Design and construct a small-scale plant for removing yeast cells and a cloudy protein 'haze' from the beer. (Your materials could be supported in an upturned plastic 2 dm³ lemonade bottle from which the base has been cut away.)